INDWELLING NEURAL IMPLANTS
STRATEGIES for CONTENDING with the
IN VIVO ENVIRONMENT

FRONTIERS IN NEUROENGINEERING

Series Editors
Miguel A.L. Nicolelis, M.D., Ph.D.
Sidney A. Simon, Ph.D.

Published Titles

Indwelling Neural Implants: Strategies for Contending with the *In Vivo* Environment
William M. Reichert, Duke University, Durham, North Carolina

Electrochemical Methods for Neuroscience
Adrian C. Michael, University of Pittsburg, Pennsylvania
Laura M. Borland, University of Pittsburg, Pennsylvania

INDWELLING NEURAL IMPLANTS
STRATEGIES for CONTENDING with the
IN VIVO ENVIRONMENT

Edited by
William M. Reichert
Duke University
North Carolina

CRC Press
Taylor & Francis Group
Boca Raton London New York

CRC Press is an imprint of the
Taylor & Francis Group, an **informa** business

CRC Press
Taylor & Francis Group
6000 Broken Sound Parkway NW, Suite 300
Boca Raton, FL 33487-2742

First issued in paperback 2019

ISBN-13: 978-0-8493-9362-4 (hbk)
ISBN-13: 978-0-367-38793-8 (bk)

Library of Congress Cataloging-in-Publication Data

Indwelling neural implants : strategies for contending with the in vivo
    environment / editor, William M. Reichert.
        p. ; cm. -- (Frontiers in neuroengineering)
    "A CRC title."
    Includes bibliographical references and index.
    ISBN 978-0-8493-9362-4 (hardbook : alk. paper)
        1. Implants, Artificial. 2. Brain. 3. Electrodes. 4. Foreign-body reaction. 5.
Nerve tissue. 6. Wound healing. I. Reichert, William M. II. Series: Frontiers in
neuroengineering (Boca Raton, Fla.)
        [DNLM: 1. Central Nervous System--physiopathology. 2. Wound
Healing--physiology. 3. Electrodes, Implanted--adverse effects. 4. Foreign-Body
Reaction--pathology. WL 300 I421 2007]
    RD594.I515 2007
    617.9'5--dc22   2007029372

Visit the Taylor & Francis Web site at
http://www.taylorandfrancis.com

and the CRC Press Web site at
http://www.crcpress.com

To Shirley Sue Reichert.
Mother. Advisor. Friend.

Contents

PART I

PART II

PART III

PART IV

Series Preface

The Frontiers in Neuroengineering series presents the insights of experts on emerging experimental techniques and theoretical concepts that are or will be at the vanguard of neuroscience. Books in the series cover topics ranging from methods to investigate apoptosis to modern techniques for neural ensemble recordings in behaving animals. The series also covers new and exciting multidisciplinary areas of brain research, such as computational neuroscience and neuroengineering, and describes breakthroughs in fields such as insect sensory neuroscience, primate audition, and biomedical engineering. The goal is for this series to be the reference that every neuroscientist uses to become acquainted with new methodologies in brain research. These books can be used by graduate students and postdoctoral fellows who are looking for guidance to begin a new line of research.

Each book is edited by an expert and consists of chapters written by the leaders in a particular field. Books are richly illustrated and contain comprehensive bibliographies. Chapters provide substantial background material relevant to the particular subject. Hence, the books in this series are not the usual type of method books. They contain detailed "tricks of the trade" and information as to where particular methods can be safely applied. In addition, they include information about where to buy equipment and Web sites helpful in solving both practical and theoretical problems. Finally, they present detailed discussions of the present knowledge of the field and where it should go.

We hope that, as the volumes become available, that our efforts as well as those of, the publisher, the book editors, and the individual authors will contribute to the further development of brain research. The extent to which we achieve this goal will be determined by the utility of these books.

Sidney A. Simon, Ph.D.
Miguel A.L. Nicolelis, M.D., Ph.D.
Series Editors

Preface

In the quarter century since the idea of using neural signals to control an external effector prosthetic first surfaced, enormous strides have been made in understanding the neuronal circuitry of the relevant brain structures, developing the computer hardware and software algorithms capable of transforming the neuronal potentials into control signals, and engineering high-density recording arrays. In contrast, a number of fundamental and unresolved questions remain as to the source of signal degradation in chronically implanted neuroelectrodes. Is it the result of insertion trauma, micromotion, mechanical mismatch, or simply the consequences of glial scar formation arising from a normal chronic foreign body response? What roles do electrode size, shape, surface chemistry, mechanical impedance, and insulating material play? What is the fate of neurons adjacent to the recording site as the signal degrades? Are they silenced, killed, or just pushed out of the way by the glial scar? Do these effects arise from inhibitory cues expressed in the glial scar or by the inflammatory molecules released into the vicinity of the electrode? What happens when one introduces devices that stimulate the surrounding brain tissue? What is the impact on the surrounding tissue from telemetric communication and on-chip processing? What are the best methods for intervening in the repair of trauma caused by device implantation? At this point, nobody really knows; however, if answers are not found, then the field will never progress from its current state of the occasional well-publicized success to widespread use in humans.

This text entitled *Indwelling Neural Implants: Strategies for Contending with the In Vivo Environment* is a compendium of contributions from a number of noted experts who are working in the field of neuroprosethetics and tissue repair. These experts were requested to contribute chapters that focused on this general theme, but I also encouraged them to elaborate on specific aspects important to future therapies for the response of central nervous system (CNS) tissue to implants. Combined, this text is a unique contribution to the CNS literature and one that certainly fits the theme of Taylor & Francis/CRC's Frontiers in Neuroengineering series.

The central feature of these chapters is the impact, characterization, and mitigation of the healing of the tissue that surrounds the implant. By and large, these chapters speak for themselves; however, the text can be divided thematically into four parts.

Part I reviews and highlights the differences between wound healing in the CNS, peripheral nervous system, subcutaneous tissue, and bone. While healing in these tissues shares a number of molecular and cellular phenomena, differences arise rapidly as one progresses from the initial trauma to inflammation to repair. This chapter serves as the phenomenological basis for the subsequent contributions.

Part II by Warren Gill and Patrick Wolf of Duke University is composed of two chapters that present the performance of implanted neuroprosthetics from the components perspective. Compared to the earliest "passive" recording electrodes, neuroprosthetics are increasingly "active" implants that impact the surrounding tissue chemically, mechanically, thermally, and electrically. These chapters present performance issues that arise from chronic tissue stimulation and heating effects from on-chip and telemetric processing.

In Part III, Jau-Shyong Hong of the National Institute of Environmental Health Sciences and Patrick Tresco of the University of Utah contribute, respectively, *in vitro* and *in vivo* approaches to assessing the CNS wound healing response to materials. Dr. Hong's chapter compares the complexity and the level of information obtained from the different cell culture and organotypic models. Dr. Tresco describes the application and numerical modeling of his chronic hollow-fiber-membrane *in vivo* implant model for assessing molecular transport to the surrounding brain tissue.

Part IV consists of three chapters that describe molecular and materials strategies for intervening in CNS wound repair and enhancing the electrical communication between the electrode surface and the surrounding tissue. The chapter by Ravi Bellamkonda of the Georgia Institute of Technology stresses molecular approaches for biasing cellular migration during the various stages of healing around implanted electrodes to encourage greater and more specific cell–surface interaction. The chapter by David Martin of the University of Michigan presents the deployment and characterization of novel conducting polymers to facilitate the communication between the electrode surface and the adjacent neurons. Finally, the chapter by Molly Shoichet of the University of Toronto and Michelle LaPlaca of the Georgia Institute of Technology takes the broader view of addressing the sequelae of CNS traumatic injury (of which neuroprosthesis insertion is a specific example) and presents novel strategies for nerve regeneration and repair.

William M. Reichert, Ph.D.
Duke University
Durham, North Carolina

Acknowledgments

I extend my sincere appreciation to the lead authors and their coauthors for the hard work in preparing their fine contributions. I also thank series editors Sidney Simon and Miguel Nicolelis of Duke University for the opportunity to compile this text.

Editor

William M. Reichert, Ph.D., was born in San Francisco, California, grew up in Ann Arbor, Michigan, and lives in Hillsborough, North Carolina, with his wife Kate, his son Stephen, and three dogs and a cat. He also has a daughter Elizabeth and a stepdaughter Miranda. He graduated with a B.A. in Biology and Chemistry from Gustavus Adolphus College in 1975, was a part-time student, a hospital technician and bartender for a couple of years, and received a doctorate in Macromolecular Science and Engineering from the University of Michigan in 1982. He was a National Institutes of Health (NIH) National Research Service Award Postdoctoral Fellow, a Whitaker Fellow, and a NIH New Investigator Fellow at the University of Utah in the Department of Bioengineering, where he "learned the ropes" from Professors Joe Andrade and Art Janata.

He joined the Department of Biomedical Engineering at Duke University in 1989 and is currently Professor of Biomedical Engineering and Chemistry, and Director of the Center for Biomolecular and Tissue Engineering. He is a fellow of the American Institute of Medical and Biological Engineering and is on the editorial boards for the *Journal of Biomedical Materials Research* and *Langmuir.* He is program director of an NIH predoctoral training grant that supports graduate fellowships in biotechnology and has been a steadfast advocate for promoting underrepresented minorities in engineering graduate education, receiving the Catalyst for Institutional Change from Quality Education for Minorities (QEM) Network, Pioneer Award from the Samuel DuBois Cook Society, and the Dean's Award for Excellence in Mentoring. His current research interests are wound healing-related implant failure, biosensors, vascular graft endothelialization, and cytokine profiling. He has trained a number of doctoral and postdoctoral students now working in academics and industry, he has published nearly 100 scientific papers, and he holds patents for multianalyte waveguide sensors and protein detection arrays.

Contributors

Mohammad Reza Abidian
Department of Biomedical Engineering
University of Michigan
Ann Arbor, Michigan

M. Douglas Baumann
Department of Chemical Engineering
    and Applied Chemistry
University of Toronto
Toronto, Ontario, Canada

Ravi V. Bellamkonda
Coulter Department of Biomedical
    Engineering
Georgia Institute of Technology/Emory
    University
Atlanta, Georgia

Michelle Block
Neuropharmacology Section
National Institute of Environmental
    Health Sciences
National Institutes of Health
Research Triangle Park, North Carolina

Michael J. Bridge
Department of Bioengineering
University of Utah
Salt Lake City, Utah

Warren M. Grill
Department of Biomedical Engineering
Duke University
Durham, North Carolina

Wei He
Coulter Department of Biomedical
    Engineering
Georgia Institute of Technology/Emory
    University
Atlanta, Georgia

Jeffrey L. Hendricks
Department of Biomedical Engineering
University of Michigan
Ann Arbor, Michigan

J. S. Hong
Neuropharmacology Section
National Institute of Environmental
    Health Sciences
National Institutes of Health
Research Triangle Park, North Carolina

Dong-Hwan Kim
Department of Biomedical Engineering
Duke University
Durham, North Carolina

Michelle C. LaPlaca
Coulter Department of Biomedical
    Engineering
Georgia Institute of Technology/Emory
    University
Atlanta, Georgia

David C. Martin
Department of Biomedical Engineering
Biomedical Engineering and
    Macromolecular Science and
    Engineering
University of Michigan
Ann Arbor, Michigan

Vadim Polikov
Department of Biomedical Engineering
Duke University
Durham, North Carolina

Laura Povlich
Macromolecular Science and
    Engineering
University of Michigan
Ann Arbor, Michigan

Sarah Richardson-Burns
Department of Biomedical Engineering
University of Michigan
Ann Arbor, Michigan

Molly S. Shoichet
Department of Chemical Engineering
    and Applied Chemistry
University of Toronto
Toronto, Ontario, Canada

Sarah Spanninga
Macromolecular Science and
    Engineering
University of Michigan
Ann Arbor, Michigan

John D. Stroncek
Department of Biomedical Engineering
Duke University
Durham, North Carolina

Ciara C. Tate
Coulter Department of Biomedical
    Engineering
Georgia Institute of Technology/Emory
    University
Atlanta, Georgia

Patrick A. Tresco
Department of Bioengineering
University of Utah
Salt Lake City, Utah

Patrick D. Wolf
Department of Biomedical Engineering
Duke University
Durham, North Carolina

Cen Zhang
Neuropharmacology Section
National Institute of Environmental
    Health Sciences
National Institutes of Health
Research Triangle Park, North Carolina

Part I

1 Overview of Wound Healing in Different Tissue Types

John D. Stroncek and W. Monty Reichert

CONTENTS

1.1 INTRODUCTION AND OVERVIEW

The inevitable response to any implant is wound healing comprised of hemostasis, inflammation, repair, and remodeling. For nondegradable smooth-surfaced implants, repair and remodeling leads to isolation of the implant through tissue encapsulation. The nature of the encapsulation tissue and the cellular participants in the immune reaction leading to this outcome varies depending on the site of implantation and the type of tissue that hosts the implant (not to mention the skill of the surgeon).

It is now well-established that the wound healing process has substantial deleterious effects on the fidelity and reliability of implanted sensors and electrodes. A number of reviews [1–6] and a multitude of primary articles address this issue for implanted sensors, primarily glucose electrodes; however, there has been only one review [7] and a limited number of primary articles regarding electrodes implanted in central nervous system (CNS) tissue [8–11]. This chapter will provide a summary of the wound healing process in CNS tissue and compare it to the wound healing process in the peripheral nervous system (PNS), bone, and skin.

Recent reports have shown that implanted electrodes can monitor neural signals in the CNS and therefore can be used for the control of prosthetics in patients suffering from full or partial paralysis [12,13]. While initial studies are promising, chronically implanted sensors and electrodes frequently and unpredictably experience component failure or complications arising from the wound healing response [7]. The wound healing process begins as soon as the electrodes are inserted into the brain, inevitably disrupting the blood–brain barrier [14]. Breaching of the blood–brain barrier leads to hemostasis and the initial stages of inflammation in CNS tissue that are typical of those seen in all vascularized tissue. However, distinct differences arise in CNS tissue during the later stages of inflammation, repair, and remodeling (Figure 1.1).

The first difference is the blood–brain barrier. The CNS is highly vascularized; however, maintaining a barrier between capillaries and the surrounding tissue is important for regulating the concentration of ions and other molecules in both the vascular and extravascular space of the CNS [15]. Endothelial cells forming capillaries in the CNS differ from those in the rest of the body in two respects. First, they are able to form tight junctions, restricting paracellular flux, and, second, they have very few endocytotic vesicles, limiting transcellular solute movement from the blood to the brain interstices (reviewed by Rubin and Staddon [16]). As a result, the blood–brain barrier impedes entry of virtually all blood molecules, except those that are small and lipophilic, such as steroids. However, some larger molecules such as glucose and certain amino acids are still able to move into brain tissue through specific transporter proteins [16]. A dense basement membrane and astrocyte processes, termed end-feet, surround capillary endothelial cells, further contributing to the blood–brain barrier. Under normal nonpathological conditions the CNS vascular endothelium with its tight junctions completely separates brain tissue from circulating blood, isolating the brain's unique cell types while protecting neural tissue from potentially harmful immune responses occurring outside of the neural tissue in the body [17]. This is important because unlike most peripheral tissues, the CNS functions through a network of neurons that is generally incapable of proliferation and cannot be replaced when impaired.

CNS tissue is distinguished from non-CNS tissue by the presence of cell types that are unique to the CNS. Neurons, which consist of a cell body and its axon, are the functional unit of brain tissue even though they account for fewer than 25% of the total cell population [7]. About half of the total CNS volume is occupied by the three types of supporting connective tissue glial cells: oligodendrocytes, astrocytes, and microglia. Each glial cell type performs functions that ensure the health of neurons in the CNS. Oligodendrocytes wrap around axons to provide support and produce a lipid-rich myelin sheath to protect neural processes; astrocytes have many neural support functions but are notable for forming the glial boundary between CNS and non-CNS structures to create the blood–brain barrier; microglia share many phenotypical and functional characteristics with blood-derived macrophages,

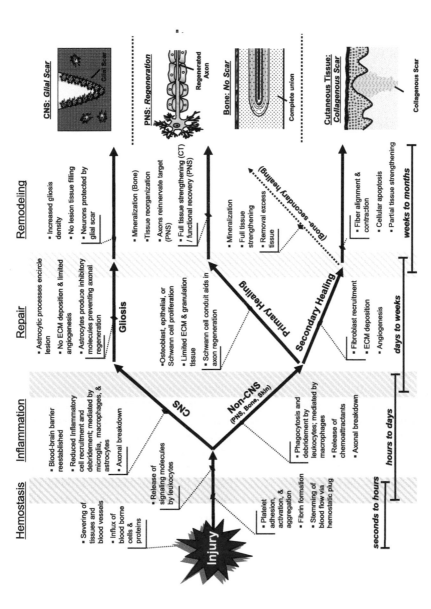

FIGURE 1.1 Summary of wound healing response in the brain versus nonneural tissue. **(See color insert following page 110.)**

existing in a resting state within brain tissue until activated to marshal the innate immune responses of the brain [7,18]. In contrast to the CNS, axons in the PNS have a continuous myelin sheath produced by Schwann cells. The continuous myelin sheath increases the propagation velocity of the nerve impulse, which is important for the axons that extend long distances through the PNS.

Subsequent sections of this chapter highlight the characteristics of each stage of the wound healing response (i.e., hemostasis, inflammation, repair, remodeling) while identifying the similarities and differences in the main tissue types. However, it is important to note that there are no defined boundaries between stages as the wound healing response "transitions" into the next stage of healing.

1.2   HEMOSTASIS (SECONDS TO HOURS)

Common to all tissue types, the wound healing process begins immediately after injury as tissue is disrupted and blood vessels are severed, releasing blood plasma and peripheral blood cells into the wound site. The earliest signals of tissue injury are the release of molecules such as ATP and the exposure of collagen on the blood vessel wall [19]. In the CNS, traumatic brain injury ruptures the blood–brain barrier as serum and blood cells are released into the wound site [14]. In all tissues, a clot is formed that acts as a temporary barrier that prevents excess bleeding and limits the spread of pathogens into the blood stream. Primary hemostasis occurs as platelets adhere to collagen fibers exposed in the damaged endothelium using specific collagen receptor glycoproteins (GPIb/IX/V) to form the primary hemostatic plug [19]. As platelets attach to the lesion site, they rapidly upregulate the high-affinity platelet integrin αIIbβ3, which mediates platelet aggregation [19]. Once platelets bind, they activate and degranulate, releasing their contents into the plasma to stimulate local activation of plasma coagulation factors (Figure 1.2b). These factors trigger the generation of a fibrin clot (Figure 1.2c) [20]. Exposure of blood plasma to tissue factor, produced by subendothelial cells not normally exposed to blood, such as smooth muscle cells and fibroblasts [21], or to foreign surfaces, such as implants, initiates an accelerated cascade of activated proteins that leads to fibrin formation. The cleaving of fibrinogen to fibrin monomers and its polymerization and cross linking forms an intertwined gelatin-like platelet plug, producing a stable clot [22].

Platelet activation also causes the release of a number of signaling molecules such as platelet-derived growth factor (PDGF) [23], transforming growth factor-β (TGF-β) [24], and vascular endothelial growth factor (VEGF) [25] from their cytoplasmic granules. Inflammatory and reparative cells are chemotactically attracted to the "reservoir" of molecules stored within the clot that gives rise to inflammation, the next step in the sequence of healing.

1.3   INFLAMMATION (HOURS TO DAYS)

1.3.1   INITIAL EVENTS

The process of inflammation contains, neutralizes, or dilutes the injury-causing agent or lesion, regardless of tissue type [26]. In both the CNS and non-CNS, the tissue environment in which inflammation begins is a mixture of injured tissue,

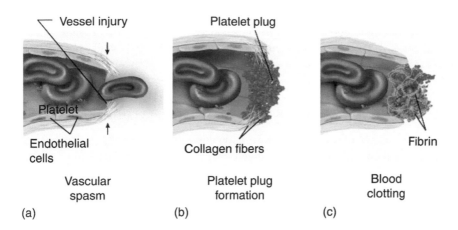

FIGURE 1.2 Hemostasis in damaged blood vessel. (a) An injury to the vessel wall causes the release of blood-borne cells, proteins, and platelets into the periphery. Platelets are activated upon coming into contact with the surrounding collagen outside of the blood vessel. (b) Activated platelets adhere together, forming an initial plug to stem blood loss. Once bound, they degranulate, releasing their contents into the plasma. (c) Meanwhile the coagulation cascade is initiated, which results in the production of fibrin to reinforce the clot. (Reprinted from Saladin, K. Anatomy and Physiology, McGraw-Hill, 2004. With permission.)

components of the clot (platelets, erythrocytes, and fibrin), extravasated serum proteins, and foreign material introduced at the time of the injury [27]. Inflammation is initiated by the release of signaling molecules from the wound site during hemostasis [28]. Chemoattractant molecules released by platelets also increase vasodilation and vascular permeability, subsequently enhancing leukocyte recruitment. Incoming leukocytes recognize plasma proteins such as fibronectin, vitronectin, and thrombospondin, which are passively absorbed by the clot. The initial process of inflammation is common to all tissue types because any physical trauma will usually result in the formation of the platelet plug and the recruitment of blood-borne inflammatory cells to the injury site. In CNS tissue, injury also results in the activation of microglia.

Neutrophils are the first inflammatory cells recruited to the wound, and they arrive within 24 hours after injury. They migrate into the wound by responding to chemoattractants released by platelets as well as chemokines presented on the endothelial cell surface [29]. The leukocyte receptor PSGL-1 binds to P-selectin expressed on both platelets and endothelial cells [30]. The low-affinity selectin binding to PSGL-1 on the neutrophil membrane causes flowing neutrophils to roll and briefly tether to endothelial cells. During rolling, neutrophils are activated further by interleukin-8 (IL-8) and macrophage inflammatory protein (MIP-1β) released from the endothelial cells. Neutrophil receptor binding to chemoattracants activates their β2 integrins, which then firmly attach to endothelial cell intracellular adhesion molecules (ICAM)-1 [31]. Neutrophils subsequently extravasate through the vessel wall into the wound site, where they release proteolytic enzymes for the digestion of foreign debris and the killing of bacteria through phagocytosis and superoxide and hydrogen peroxide production [32]. Neutrophils undergo apoptosis after 24 to

48 hours if wound decontamination is complete. They are replaced by extravasating monocytes and macrophages brought to the site of injury by further release of MIP-1α and MIP-1β, monocyte chemoattractant protein-1 (MCP-1), and a chemotactic cytokine called RANTES, all of which are produced by activated endothelial cells [33].

1.3.2   LATER EVENTS (CENTRAL NERVOUS SYSTEM [CNS] VERSUS NON-CNS)

Within 2 to 3 days after injury, inflammation continues as monocytes are recruited from the blood and differentiate into macrophages. Once in the tissue, macrophages release additional proinflammatory cytokines such as IL-1, TGF-β, and tumor necrosis factor-α (TNF-α) [26]. Macrophages remove foreign debris and can remain present for as long as a few months, depending on the extent of injury and the amount of foreign and necrotic debris to be cleared [26]. Macrophages are believed to be more important than neutrophils for successful inflammation resolution. In studies where neutrophils were depleted, wound repair was not disturbed [34], but when macrophages were removed instead, there was limited clearance of necrotic debris at the wound site, resulting in a modified wound healing process [32,25]. Mast cells within the tissue contribute to the inflammatory response by releasing histamine and serotonin to enhance blood vessel permeability and promote macrophage migration.

Once the blood–brain barrier is reestablished through the patching of damaged blood vessels during the later stages of hemostasis, the inflammatory responses of the CNS and non-CNS tissues begin to diverge (Figure 1.1). Specifically, the selectins expressed on vascular endothelial cells that attract leukocytes are less prevalent in CNS tissue, thus limiting recruitment [36–38]. Consequently, axon and myelin debris from damaged neurons can take months to clear in the CNS while only taking days in the PNS [39]. Reduced leukocyte activity and infiltration can be seen as a method to protect the brain from undue inflammatory damage, with the decreased levels of leukocyte recruitment replaced by the resident macrophage-like microglia cells.

Microglia, the only resident phagocytoic cells in the CNS, are normally found in a highly branched resting or "ramified state" and only become activated if debris needs to be cleared or if the blood–brain barrier is compromised [7]. Upon activation, microglia withdraw their processes and change from their resting ramified state to a more compact rod-like shape [40,41]. Microglia act in concert with macrophages that have penetrated the blood–brain barrier to phagocytose degenerating axons and myelin at the site of injury [42]. At this point, microglia can be distinguished from blood-derived macrophages through flow cytometric analyses *in vitro* by differences in CD45 expression (microglia: CD45 low; macrophages: CD45 high) [43,44].

Microglia, like macrophages, express surface receptors involved in the innate immune response after exposure to various pathogens or pathogen products such as lipopolysaccharide (LPS) [45]. Activated microglia also contribute aspects of adaptive immunity by upregulating major histocompatibility complex class II and costimulatory molecules to present antigens to incoming T-cells [45,46]. Microglia have been shown to appear stimulated within 1 hour after an initial tissue injury

[10,47]. The number of activated microglia peaks at 3 days after an inflammatory insult, although microglia can remain activated up to 1 or 2 weeks after injury [48].

Microglia at the site of injury produce a variety of proinflammatory and neuro-toxic factors such as MCP-1 [49], TNF-α [50], IL-1β [51], IL-6 [52], nitric oxide [53], and superoxide [54]. Paradoxically, microglia have also been reported to increase neuronal survival through the release of antiinflammatory factors, such as IL-10 [55,56], and neurotrophic factors, such as nerve growth factor (NGF), brain-derived neurotrophic factor (BDNF), and neurotrophin-3 (NT-3) [57,58] (reviewed in [58]).

Astrocytes, which become activated within hours after injury, are the second type of effector cells in the immune response that reside in the CNS. Astrocytes also perform a wide variety of functions outside of inflammation such as providing growth cues and mechanical support for neurons during CNS development, creating and maintaining the blood–brain barrier, and helping control the surrounding chemical environment through the buffering of neurotransmitters and ions released during neuronal signaling [7]. Activated astrocytes are identified during inflammation by upregulation of glial fibrillary acid protein (GFAP), increased proliferation, and hypertrophic cytoplasm and nuclei. GFAP is upregulated as early as 1 hour post injury and is thus a sensitive early marker of reactive astrocytes [59]. However, GFAP upregulation is not solely indicative of astrocyte activation since microglia can engulf necrotic astrocytic GFAP fragments and thereby express the astroglial marker [60]. Therefore, cellular hypertrophy is an important co-indicator of astrogliosis. Astrocytes have been shown to help mediate inflammation through the release of both proinflammatory and antiinflammatory factors such as: (a) nitric oxide, which contributes to the innate immune response while also causing neuro-degeneration; (b) expression of both class I and class II MHC molecules that aid in the secondary immune response; and (c) regulation of blood vessel diameter through release of vasoconstrictors [60].

Three days into an inflammatory response in the CNS, activated astrocytes are observed at the periphery of injury while macrophages and microglia are colocalized at the center of the wound [48] (Figure 1.3A). Microglia are likely the major players in the response since macrophages are largely excluded from the injury by the blood–brain barrier [61]. Nevertheless, macrophages and lymphocytes from nonneural tissue can also enter into the brain late in the inflammatory stage to further amplify the body's wound healing response by secreting a variety of cytokines. However, only lymphocytes that are in an activated state are able to pass through the blood-brain barrier and enter the CNS (for review see [62]).

The mechanism of lymphocyte extravasation during inflammation is modified in the CNS. First, there is no lymphocyte rolling along the epithelium within CNS, and, second, an activated lymphocyte can extravasate directly through an endothelial cell, leaving the neighboring tight junctions intact to preserve the blood–brain barrier [63]. This process can take several hours, whereas diapedesis through blood vessels into lymph nodes outside the CNS can occur within minutes [64]. The lymphocyte entry level varies within tissues of the CNS, with the highest entry in the spinal cord and the lowest entry in the cerebrum [36]. However, T cells find the CNS environment hostile and die rapidly via apoptosis or leave the CNS via the cere-

A CNS

B Partial-Thickness Wound to Cutaneous Tissue

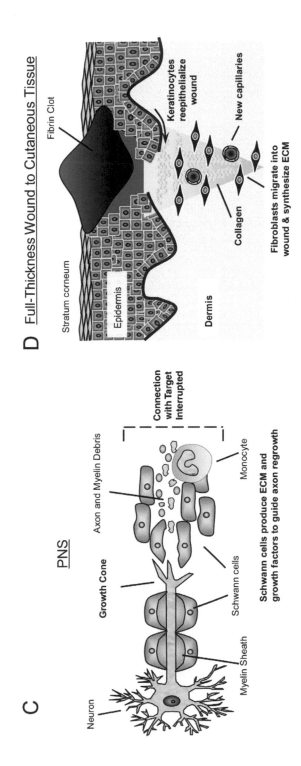

FIGURE 1.3 Wound repair of different tissue types. (A) In the CNS, microglia and macrophages migrate into the lesion, secreting various cytokines and growth factors while also removing necrotic tissue. Activated astrocytes form a physical barrier around the area of tissue damage, protecting the residential neurons outside the wound from further insult. Inhibitory molecules associated with myelin and produced by astrocytes within the glial scar block regeneration of damaged axons. (B) In partial thickness wounds, where only the epidermis is damaged and the basement membrane remains intact, the wound can heal by keratinocyte regeneration and migration alone. (C) Following a cut in the PNS, monocytes and macrophages migrate into the lesion to phagocytose axon and myelin debris. Schwann cells distal to the injury proliferate, forming a conduit for facilitating in axonal regrowth. Axons sprout and regenerate through the Schwann cell scaffold in the attempt to reinnervate the damaged tissue. (D) Repair of the lower dermal layer occurs via extracellular matrix (ECM) production by fibroblasts coupled with the formation of granulation tissue as endothelial cells produce new capillaries. The upper epidermal layer is repaired as keratinocytes produce proteases to tunnel through the fibrin clot and over granulation tissue in the attempt to re-establish the damaged basement membrane. **(See color insert.)**

brospinal fluid [36,65]. The reduced lymphocyte level results in significantly lower levels of immune surveillance in the CNS than in other tissues and is a reason why the CNS is often referred to as a site of "immunological privilege" [36]. A consequence of the inflammatory response in the brain is neuronal damage, as evidenced by the absence of neurofilaments around the lesion only a few days after injection of a proinflammatory agent [48].

The inflammatory phase is resolved by a mixture of antiinflammatory cytokines, although the specific individual cytokines that are involved and the mechanisms of their actions are largely unknown [66]. TGF-β secreted by macrophages is considered one of the main attenuators of inflammation in both CNS and non-CNS tissue [66]. TGF-β is involved in promoting the recruitment and proliferation of fibroblasts to push the wound healing response toward the repair phase. Macrophages are also known to produce other prowound healing factors such as VEGF to initiate angiogenesis within the hypoxic tissue environment [67].

1.4   REPAIR (DAYS TO WEEKS)

During inflammation the overall tissue strength of a wound is minimal, since tissues do not regain their normal functional strength until inflammation transitions into repair. This transition is mediated by macrophages and their antiinflammatory cytokines and generally occurs one week after the initial injury. The end result of the repair process is vitally important to wound healing because it establishes the scaffolding necessary to support and rebuild the damaged tissue.

Tissue repair is characterized by increased cell proliferation, capillary budding, and the synthesis of extracellular matrix (ECM) to fill in the damaged tissue that has been cleared during inflammation. The matrix material is initially made up of fibrinogen and fibronectin [68]. Thereafter, proteoglycans, large macromolecules with a core protein and one or more covalently attached glycosaminoglycan molecules, are synthesized by cells to make up the ground substance of the ECM. ECM-producing cells that produce the support matrix necessary to regain structural integrity include fibroblasts in connective tissue, chondrocytes in cartilage, osteoblasts in bone, and Schwann cells in the PNS.

The body's reparative response to injury differs between soft collagenous cutaneous tissue, bone, the PNS, and the CNS (Figure 1.1). While complete repair of an injured tissue is ideal, this response can only consistently occur in mineralized tissue or bone, as re-establishing complete structural integrity is critical for functional recovery. In unmineralized connective tissue, specifically skin wounds, the depth of tissue loss dictates the repair response. While the epidermis can regenerate, deep tissue wounds in which the dermis is lost undergo secondary healing that requires excess ECM production, leading to the formation of fibrous scar tissue. In the right conditions, regeneration can occur in the PNS as injured axons can regrow through a Schwann cell scaffold. In the CNS, lost axons are not replaced, and a glial scar is formed by reactive astrocytes acting as a physical and molecular barrier that inhibits axon regeneration.

1.4.1 NON-CNS TISSUE

The first stage of tissue repair is stabilization of the discontinuity created by the injury. Traditionally, there are two broad classifications of healing. Tissue that has little to no gap separating the wound boundaries will undergo "primary healing" from the apposed edges of the tissue. Tissue that is unstable with a large gap or discontinuity injury will undergo "secondary healing," where excess ECM is produced to secure and fill the lesion. The ECM of secondary healing, which subsequently becomes vascularized, is referred to as granulation tissue—a term arising from its appearance. In general, the amount of granulation tissue formed is proportional to the eventual level of scarring.

1.4.1.1 Primary Repair (No Extracellular Matrix [ECM] Production)

1.4.1.1.1 Partial-Thickness Skin Repair

Skin is composed of two distinct layers, the epidermis and the dermis. In partial thickness epidermal wounds, only the epidermis is damaged, leaving the basement membrane intact along with hair follicles and sweat and sebaceous glands. Because only the epidermal surface needs to be replaced and epithelial progenitor cells remain intact below the wound, the synthesis and deposition of collagen is not required. To repair a surface lesion such as a "paper cut," the site must only be reepithelialized by migration of keratinocytes from below the wound and at the wound edge (Figure 1.3B). The degree of reepithelialization depends on the amount of tissue loss and the depth and width of the wound.

Reepithelialization begins within 24 to 48 hours as uninjured keratinocytes detach from the basal lamina to bore through or underneath the fibrin clot, crawling into and across the wound—a process termed lamellopoidial crawling (for review see [69]). Migration starts as keratinocytes at the wound edge upregulate their production of matrix metalloproteinases (MMPs), releasing the cells from their tethers to the basal lamina [70]. Keratinocytes resting on the basal lamina migrate across the wound site at a rate of 1 to 2 mm/day [71], attaching to fibronectin and vitronectin contained within the clot by upregulating their expression of $\alpha_5\beta_1$ and $\alpha_v\beta_6$ integrins [69,72]. While moving through the dense fibrin clot, keratinocytes dissolve the dense fibrin matrix through the upregulation of several proteases such as tissue plasminogen activator (tPA) and urokinase-type plasminogen activator (uPA), activating the fibrinolytic enzyme plasmin [73].

Keratinocyte locomotion occurs through the contraction of actinomyosin filaments of the cytoskeleton [69]. Because of changes in integrin expression and actinomyosin filament assembly as keratinocytes attach to the clot's provisional matrix, there is a delay of several hours before the migration of basal keratinocytes is observed [74]. Once migration begins, keratinocytes move one by one over the wound site until the wound is covered by a complete layer of keratinocytes [72]. After migration is complete, the cells behind the wound edge undergo a proliferative burst to replace keratinocytes lost resulting from injury while also forming additional keratinocyte layers over the basal layer [75]. It is believed that epidermal growth factor (EGF), keratinocyte growth factor (KGF), and transforming growth factor (TGF-α) may drive cell proliferation and wound closure [69]. Migration and proliferation continue until keratinocytes receive a stop signal, likely due to contact inhibition [28]. At this time, MMP expression is interrupted, and a new basement membrane is produced whereby new cell-matrix adhesions are established [28].

1.4.1.1.2   Stabilized Bone Repair

When motion at the fracture site is prevented and bone fractured ends are rigidly held in place immediately after injury, primary mineralized tissue repair occurs. In primary bone repair there is no need for ECM-rich structural support because the tissue is already stabilized. Two types of primary mineralized tissue healing can occur: gap repair, where there is a less than 0.5-mm breach separating stabilized bone fragments, and contact repair, where the fractured ends are held in direct apposition [76].

In gap repair, healing begins as blood vessels and loose connective tissue fill the wound. After 2 weeks pluripotent mesenchymal cells derived from the bone marrow arrive at the site of injury and differentiate into bone-producing cells called osteoblasts. Osteoblasts fill any gaps in the tissue by secreting layers of unmineralized bone matrix called osteoid at a rate of 1 to 2 µm per day, which becomes mineralized after approximately 10 days [77]. The new bone formed in this stage is perpendicular to the long axis of the bone (Figure 1.4A). As osteoid becomes mineralized, some osteoblasts remain embedded in the matrix and become entrapped as the tissue is mineralized. These osteoblasts become osteocytes and are responsible for mechanotransduction within the tissue. Osteoblasts are likely activated by proteins belonging to the TGF-β family such as bone morphogenic proteins (BMPs) [78]. The new bone fills the gap but does not completely unite the fracture ends. At this point, the interface between the new and original bone is the weakest link in the union [79]. This newly formed bone acts as a scaffold for future remodeling by osteoclasts and osteoblasts.

Contact repair occurs when there is direct contact between bone ends. This allows organized lamellar bone to form directly across the fracture line. However, before new bone can be deposited to repair the wound, necrotic bone must be removed. Osteoclasts, large multinucleated cells that derive from hemopoietic progenitor cells in the bone marrow and are closely related to macrophages, tunnel across the fracture line and remove old bone and necrotic tissue through the release of proteases and lysosomal enzymes (Figure 1.4B). They attach to the tissue matrix through β3 integrins, creating a ruffled cell border between the cell and the bone surface [80,81]. Hydrogen ions are pumped toward the osteoclast membrane to create an acidic microenvironment to digest the bone's mineral component. The bone's organic matrix is degraded by lysosomal proteases and is then endocytosed by osteoclasts [82]. The rate of bone degradation by osteoclasts depends on the orientation of the tissue. Osteoclasts are capable of cutting bone at a speed of 20 to 40 µm per day parallel and 5 to 10 µm per day perpendicular to the bone's long axis [83]. Capillaries form within these "cutting cones," providing nutrients to the tissue being repaired [84]. Behind the cutting cone, rows of osteoblasts line the resorptive channel, depositing layers of osteoid. This region is referred to as the "closing cone," where osteoblasts attempt to replace approximately as much bone as had been removed [83] (Figure 1.4B). Layers of osteoid are initially made up of

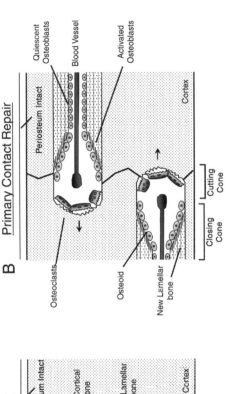

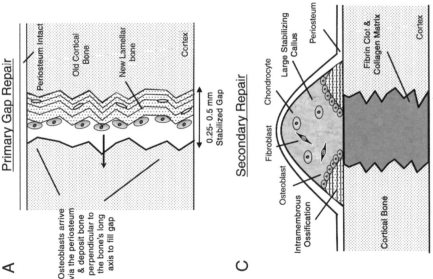

FIGURE 1.4 Primary versus secondary mineralized tissue repair. (A) When a small fracture gap occurs and the bone remains stabilized, osteoblasts deposit bone perpendicular to the bone's long axis, filling the gap and serving as a scaffold for future remodeling. (B) In primary contact repair, fractured ends are held in direct apposition. No excess ECM is required. Cutting cones are formed as osteoblasts absorb damaged bone, followed by osteoblasts synthesizing osteoid, which becomes mineralized to form lamellar bone. (C) Unstabilized fractures are initially repaired through ECM production by fibroblasts and chondrocytes arriving from the periosteum to form a stabilizing soft callus. The collagenous ECM becomes mineralized, forming woven bone, also referred to as hard callus.

collagen types I and IV, gradually becoming mineralized through the deposition of hydroxyapatite crystals to form lamellar bone. Bone formed by this method, in which there is no cartilage formation as an intermediate step, is called intramembranous ossification.

1.4.1.1.3 Peripheral Nervous System (PNS) Repair

A nerve is made up of a collection of several neurons and their supporting cells called Schwann cells (Figure 1.5). Schwann cells act as supporting cells, surrounding the signaling processes of neurons, called axons, and isolating them from the ECM by producing myelin, which increases signal propagation velocity. Each axon and its Schwann cells are encircled by a connective tissue matrix called the endoneurium, subsequently grouped into bundles, and bounded by another thin connective tissue matrix called the perineurium. A number of these bundles along with vasculature make up a nerve with an outer sheath, termed epineurium [85,86].

The possibility of repair in the PNS depends on the extent of injury. The most severe injury capable of repair is classified as a third-degree injury, where the axon and the endoneurium are damaged or severed while the perineurium and epineurium remain intact [87]. In more traumatic injuries where there is disruption of not only axons but the entire epineurium or perineurium, the blood–nerve barrier is ruptured, requiring extensive fibrous ECM to physically repair the damaged tissue. This dense ECM prevents regeneration, as the cut axon is incapable of finding its innervated target [87].

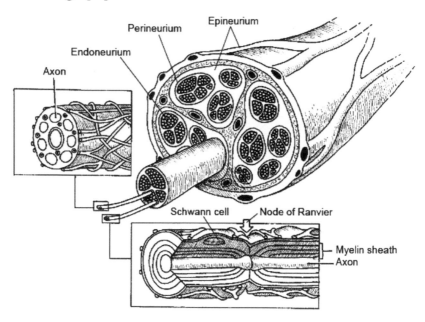

FIGURE 1.5 Cross-sectional anatomy of the peripheral nerve. Inset at left shows an unmyelinated fiber. Inset at bottom shows a myelinated fiber. (©2000 American Academy of Orthopaedic Surgeons. Reprinted from *J. Am. Acad. Orthop. Surg.*, 8(4), 243, 2000. With permission.)

No matter the extent of injury, if repair in the PNS is to be successful, the cell body of the neuron must first survive the injury to its extending axon [39]. The neuron cell body is required for metabolism in the neuron's axonal processes; therefore, any extending axon that is severed from its body quickly degenerates downstream from where the injury occurred. Within minutes to hours after injury to a neuron and its surrounding environment, degeneration begins as Schwann cells surrounding the axon stop making myelin proteins. Within the axon, enzymatic proteases are activated in response to a calcium influx that causes axon fragmentation and causes the surrounding myelin to form droplet-shaped particulate, a process termed Wallerian degeneration [39]. By 3 to 4 days, impulse conduction is no longer possible, as the extending axon is destroyed. Axonal and myelin debris is cleared during the later inflammatory stages and during the beginning of the repair phase by the influx of macrophages over a matter of weeks, in contrast to that of the CNS, which proceeds over several months [61].

In injuries where the neuron survives and the greatest extent of injury to the extending nerve is crushing or severing of the axon and the endoneurium, repair begins as Schwann cells proximal to the injury proliferate (Figure 1.3C). Schwann cells help remove the degenerated axonal and myelin debris and fill the area previously occupied by the extending axon and the myelin sheath (Figure 1.3C) [88]. The proliferating Schwann cells form interconnected cellular tubes that act as conduits for axonal regeneration [42]. Within the conduit, Schwann cells increase their production of growth factors, such as NGF and brain-derived growth factor (BDGF), while synthesizing ECM proteins such as laminin and fibronectin [39]. Axons begin to regenerate within 3 to 24 hours, sprouting multiple new small-diameter axons at the site of injury called the "growth cone," and attempt to grow through the Schwann cell conduit (Figure 1.3C) [39]. The regenerating axonal sprouts move across the lesion 0.5 to 5.0 mm per day, stimulated by contact and chemotactic guidance provided by Schwann cells [89,90]. Each axonal sprout contains filopodia that bind through $\beta1$ integrins to ECM molecules such as laminin and fibronectin that are produced by Schwann cells [91]. The newly regenerated axons are initially unmyelinated but are mitogenic, inducing adjacent Schwann cells to proliferate further and produce myelin around the axon as it regenerates [39]. Axonal sprouts continue to regrow in the attempt to reinnervate the tissue that was deinnervated by injury-induced axonal degeneration.

1.4.1.2 Secondary Repair (ECM Production)

1.4.1.2.2 Full-Thickness Cutaneous Tissue Repair

If the basement membrane is damaged in a full thickness skin wound and substantial dermis is lost, the wound cannot heal by reepithelialization alone. ECM-producing dermal fibroblasts adjacent to the wound site are activated and proliferate, migrating into the wound hematoma within 3 to 4 days [69].

In secondary repair, during approximately the same time period that keratinocytes attempt to re-epithelialize the wound (as previously illustrated in primary skin repair), fibroblasts migrate through the provisional matrix by contraction of their actinomyosin cytoskeleton. Migrating fibroblasts initially synthesize fibronectin [92], but in response to TGF-β1 released by macrophages, fibroblasts switch to a more fibrotoic phenotype, synthesizing collagen matrix that begins to provide structural support for the wound [93]. Keratinocytes utilize this newly produced ECM to migrate over and epithelialize the site of injury (Figure 1.3D). ECM acts also as a conduit where new capillaries are formed as endothelial cells migrate into the wound site, responding to growth factors such as fibroblast growth factor-2 (FGF-2) and VEGF [94]. Neovascularization delivers nutrients to the migrating fibroblasts at the site of injury, giving the replacement tissue a pink, granular appearance—hence the name granulation tissue. New blood vessels are formed as there is a shift in the balance between the relative amounts of molecules that induce versus the molecules that inhibit vascularization. Proangiogenic growth factors such as VEGF are secreted predominantly by keratinocytes on the wound edge while FGF-2 and TNF-α are released from damaged endothelial cells and macrophages [69]. Hyaluronan, an oligosaccharide component of the ECM, also promotes angiogenesis and aids in repair [95]. Revascularization of the wound site is critical, as angiogenic failure can result in chronic wounds such as venous ulcers that are unable to heal.

1.4.1.2.2   Bone Repair

For bone repair, secondary healing is more commonly seen than primary fracture healing bone because most bones are not rigidly supported after injury [96]. When injury results in the disruption of the bone's external vascular covering (periosteum) as well as the surrounding soft tissue, the damaged tissue is usually left unsupported. Unstabilized mineralized tissue will undergo secondary repair where the first step is creating an ECM-rich bridge to support the fracture (Figure 1.4C).

Most compact bone surfaces that make up the outermost layer of mineralized tissue are covered with a lining of osteoblasts that become active and produce a small amount of intramembranous ossification as early as 24 hours after injury along either side of the fracture (Figure 1.4C) [96]. However, this limited early bone formation provides little stability [96]. The wound site does not begin to regain mechanical strength until 3 to 4 days after injury, when fibroblasts and undifferentiated mesenchymal cells arrive at the periosteum via the circulation [97]. In secondary healing osteoprogenitors arriving from the vasculature differentiate into chondrocytes, whereas in primary healing they differentiate into osteoblasts. Chondrocytes and fibroblasts team to produce collagenous and fibrous tissue that forms around the outside of the fracture at the periosteum and internally within the marrow to provide an internal splint (Figure 1.4C) [98]. This collagen-rich tissue is called soft callus and can be classified as a type of granulation tissue that is critical for providing vascularity and structural support to the fracture. Maximum callus is usually observed a week after injury (Figure 1.6, day 7) [99]. In general, the amount of soft callus formation is dependent on the relative stability of the fracture fragments. The greater the motion at a fracture site, the more callus is required to prevent this motion [100]. Growth factors such as PDGF, TGF-β, and FGF released from platelets immediately after injury are possible initiators of callus formation [100].

At 2 weeks, the collagenous soft callus is gradually mineralized to form hard callus, increasing the stability of the fracture site (Figure 1.6, day 14) [99]. Colla-

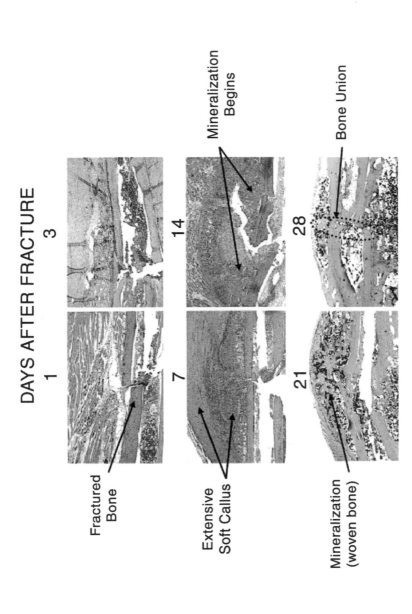

FIGURE 1.6 Histological analysis of secondary fracture healing in bone showing the progression of repair on days 1, 3, 14, 21, and 28. Fractured bone appears denser than the surrounding tissue. On day 7, extensive soft callus is seen forming around the injured bone. At day 14, the soft callus becomes mineralized to form new bone and achieve union by day 21 and 28 (H&E stain, x40). (Adapted from *J. Bone Miner. Res.*, 16, 1004–1014, 2001. With permission of the American Society for Bone and Mineral Research.)

gen mineralization to form woven bone is different from osteoid mineralization to form lamellar bone. In collagen mineralization, chondrocytes stop their production of collagen, elongate, and release proteases [101]. Glycosaminoglycans within the collagen matrix inhibit mineralization and must be removed by chondrocyte proteoglycanases for mineralization to occur [101]. As the cartilage matrix is degraded, chondrocytes differentiate and secrete angiogenic factors such as VEGF to induce capillary ingrowth from adjacent tissue [102]. Osteoblasts lining the bone surface then secrete collagen-free organic matrixes such as osteonectin and osteopontin, providing nucleation sites for the initiation of nanocrystaline calcium phosphate mineralization. Bone formed by this method, where collagen is mineralized to form bone, is termed endochondral ossification.

By the third week, the majority of the cartilage has become bone and union is achieved (Figure 1.6, day 21) [99]. At this point, the healing bone is generally able to support loads. However, even after stabilization, the newly formed bone is still weaker than normal uninjured bone. Only after the remodeling phase in which woven bone becomes remodeled to more compact laminar bone, does the tissue achieve full strength [98].

1.4.2 CNS REPAIR (GLIOSIS)

The repair of a wound in the CNS is not followed by neuronal regeneration. Unlike the response seen in the PNS, where degenerated axons can regenerate, damaged axons of the CNS initially sprout, but regeneration is impeded as the growth cones collapse within a day. CNS tissue repair begins hours after injury as astrocytes outside the lesion core and the surrounding area are activated. This marked glial response, commonly called gliosis, is made up of a multilayered sheet of activated astrocytes that form a boundary around areas of tissue damage (Figure 1.3A). Like the response seen in unmineralized tissue repair, astrocytes alter their integrin expression, migrating toward the lesion while also secreting MMPs for ECM degradation. MMP expression by astrocytes and neurons 1 to 2 weeks after injury can also promote repair by stimulating VEGF production to initiate angiogenesis [103]. Once astrocytes arrive at the site of injury, their processes encircle the lesion and become tightly intertwined to give the glial barrier a highly disordered appearance, forming what is referred to as the glial scar. The overall magnitude of glial activation roughly correlates with the amount of blood–brain barrier disruption and tissue damage [104].

Inside the lesion and surrounded by reactive astrocytes, microglia and macrophages persist in the attempt to remove potential pathogens and digest the fibrin clot [105] (Figure 1.3A). Because of the robust inflammatory response produced by microglia and macrophages, there are generally no neurofilaments at the wound site 3 days after injury [48,105]. Without neuronal viability, the CNS loses its functionality at the site of injury. From a structural standpoint, while smaller lesions in the CNS can be filled by reactive astrocytes [48], glial hypertrophy and proliferation cannot compensate for larger amounts of tissue loss. These large wounds remain cavities with reactive astrocytes forming a dense barrier around the lesion. Since the lesion is not filled with cells or ECM, surgeons can visualize past traumatic brain injuries during imaging because of missing tissue architecture (Figure 1.7A)

[106]. In cases when the outer meningeal surface of the brain is penetrated, the lesion is filled with fibrosis tissue as meningeal fibroblasts lining the outside of the brain are able to migrate into the lesion [9,105,107]. Meningeal fibroblasts, like those in connective tissue repair, are capable of producing collagen (types I, III, IV) and ECM proteins (laminin, fibronectin) to fill the wound site [108,109]. Other nonglial cells such as vascular endothelial cells and mesenchymal cells are present in the glial scar. Endothelial cells attempt to form new blood vessels, whereas mesenchymal cells deposit basal lamina, which is known to inhibit axon regrowth [60,110].

1.4.2.1   Purpose of the Glial Scar

Spinal cord injury studies where astrocytes have been inactivated have led to a better understanding of the purpose of the glial scar. Selective removal of reactive astrocytes shows no glial scar formation at 2 weeks after stab injuries [109]. However, astrocyte removal resulted in the loss of injury containment as inflammatory cells spilled into the tissues surrounding the initial wound, increasing neuronal degeneration next to the injury [109]. The glial scar is believed to protect neuronal function following injury by repairing the blood–brain barrier and to subsequently limit inflammatory response to cells of the CNS [60,104].

Unlike the PNS, axons within the CNS are unable to regenerate. Within the glial scar environment two main groups of inhibitory molecules impede axonal regeneration: those associated with the glial scar and those associated with myelin [111]. Inside the glial scar, chondroitin sulfate proteoglycans (CSPGs) are upregulated by astrocytes, oligodendrocyte precursors, and meningeal fibroblasts. CPSGs are made up of a core protein to which a variable number of repeating disaccharide chondroitin sulfate chains attach [111]. CSPGs are secreted by reactive astrocytes within 24 hours after injury and can continue for months thereafter [104,112,113]. Proteoglycan expression is highest in the center of the lesion and diminishes outward [114].

Myelin within the CNS also contains growth inhibitory ligands that are released locally following axonal trauma. These inhibitory molecules are Nogo, myelin-associated glycoprotein (MAG), oligodendrocyte-myelin glycoprotein (OMgp), Semaphorin 4D, and myelin associated CSPGs (for review see [111]). Both the glial scar and myelin-associated molecules prevent regrowth by repelling or collapsing growth cones though the Rho GTPase signaling pathway, blocking microtubule assembly [115].

The major difference between the PNS and CNS is that while the supporting Schwann cells and astrocytes both proliferate and activate after injury, Schwann cells of the PNS undergo changes to provide a supportive environment for regeneration, while astrocytes of the CNS undergo changes to produce inhibitory molecules that prevent neuronal regeneration from occurring. This has been demonstrated, as CNS axons are capable of regenerating through peripheral nerve grafts implanted into wounds outside of the CNS [116]. It has been proposed that the glial scar not only protects CNS tissue adjacent to injury from further damage but also prevents neurons from reforming inappropriate neuronal connections after injury [60,113,117].

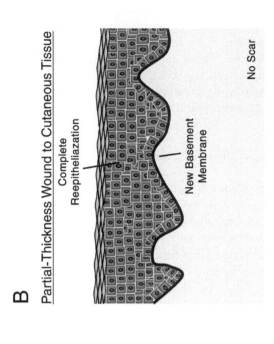

A

CNS

Activated Astrocytes   CSF-Filled Cyst   Viable Neuron

Glial Scar

B

Partial-Thickness Wound to Cutaneous Tissue

Complete
Reepitheliazation

New Basement
Membrane

No Scar

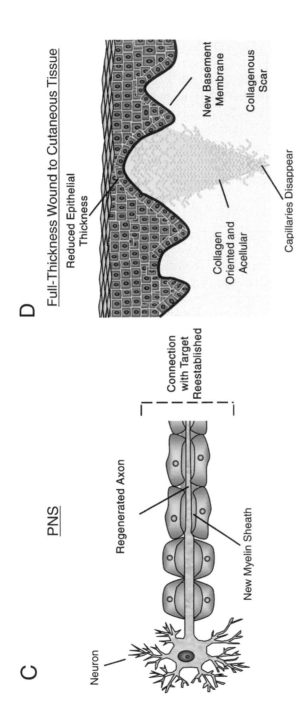

FIGURE 1.7 Wound remodeling of different tissue types. (A) At the site of injury, the fibrin clot and necrotic tissue is removed by microglia and macrophages. Unlike nonneural tissue, lost tissue is not replaced, leaving a lesion with a cerebral spinal fluid–filled cyst. Astrocytes become more dense surrounding the cyst to form a glial scar, protecting the uninjured tissue from further injury. (B) In partial thickness wounds, little or no remodeling is required because no ECM is produced during repair. (C) In the PNS, the tissue regains function as the axon successfully innervates its target while Schwann cells produce new myelin to insulate the regenerated axons. (D) Remodeling of cutaneous tissue involves contraction by fibroblasts to close the wound site, aligning the collagen matrix in response to lines of stress, and reepithelialization of the epidermis by keratinocytes. After contraction, fibroblasts within granulation tissue undergo apoptosis, leaving an acellular collagenous scar. **(See color insert.)**

Glial scarring has additionally been linked to the clearance of glutamate and the production antiinflammatory cytokines [110].

1.4.2.2   Glial Scar Induction

Cytokines such as IL-1, TGF-α, TGF-β, TNF-α, INF-α, and INF-γ have all been shown to be upregulated in scar tissue after injury, promote astrocyte proliferation *in vitro*, and augment glial scarring *in vivo* (for recent review see [104]) [110]. TGF-β1 and TGF-β2 expression has been shown to increase immediately after injury in the brain and spinal cord. TGF-β2 increased both proteoglycan production [118,119] as well as glial scarring by astrocytes [120]. INF-γ and FGF2 have also been shown to increase the extent of glial scarring and increase astrocyte proliferation in culture [104].

1.5   REMODELING (WEEKS TO MONTHS)

The ultimate endpoint following remodeling depends on the tissue type. In non-CNS tissue that undergoes primary healing, very little remodeling occurs because of the lack of ECM produced during repair. Secondary healing, in contrast, involves fiber alignment and contraction to reduce the wound size and to reestablish tissue strength. Complete recovery of original tissue strength is rarely obtained in secondary healing because repaired tissue remains less organized than noninjured tissue, which results in scar formation. Collagen-rich scars are characterized morphologically by a lack of specific organization of cellular and matrix elements that comprise the surrounding uninjured tissue. In CNS tissue where there is no repair or regeneration of injured neurons, there is also relatively little reestablishment of structural integrity in the region. Instead, during CNS remodeling, the glial scar around the lesion becomes denser as astrocytic processes become more intertwined and more or less isolates but does not repair the injured region.

1.5.1   Non-CNS Remodeling

1.5.1.1   Primary Remodeling

1.5.1.1.1 Partial-Thickness Cutaneous Tissue Remodeling

In superficial injuries, wounds can heal by epithelialization alone, with little or no additional ECM required to fill in the tissue. Because these primary tissue injuries can heal by keratinocyte regeneration and minimal ECM production, very little additional remodeling is required (Figure 1.7B). As a result, there is no scarring, and the repaired tissue is virtually indistinguishable from uninjured tissue.

In the clinical setting, when there is little contamination or necrotic debris, physicians use suturing to bring dermal edges in direct apposition. Upon careful alignment and elimination of tension, the epidermal and dermal layers can heal primarily by epithelialization within the epidermis and with limited ECM production in the dermis, leading to limited scarring [121]. Over a period of weeks to months, the wound gradually increases its tensile strength as the ECM is remodeled. During this process the injured tissue's ECM becomes reoriented with respect to tensile force. Remodeling is more critical in secondary healing and hence will be expanded upon further in the section discussing full-thickness unmineralized tissue wounds.

1.5.1.1.2   Stabilized Bone Remodeling

Mineralized tissue remodeling is an active and dynamic process. Bone is unique in that remodeling occurs throughout the life of the tissue as mechanical stress induces bone to reorient itself and produce new bone to better handle the demands that are placed on it. According to Wolff's law, the tissue adapts to the environment by becoming oriented along lines of maximum stress [112].

Remodeling after primary fracture repair where bone is stabilized is similar to the remodeling response that occurs over the life of the tissue, lasting up to several years before full preinjury strength is restored [76]. In primary gap healing, remodeling is important for restoring tissue strength. However, in primary contact healing, remodeling is coupled to the repair process. During contact remodeling, the cutting cones mature, depositing lamellar bone centripetally to form ring-shaped structures with a center blood vessel-containing canal (Figure 1.8B).

In primary gap remodeling, lamellar bone deposited perpendicular to the long axis within the injury during repair is used as a scaffold (Figure 1.8A). This process is commonly referred to as Haversian remodeling, where bone is remodeled in small packets of cells called basic multicellular units (BMUs) [123]. This is the same process that occurs during primary contact repair where osteoclasts form cutting cones allowing the influx of endothelial cells that form capillaries (Figure 1.4B). The budding capillaries bring in pluripotent mesenchymal cells that differentiate into osteoblasts after attaching to the surface of the internal cutting cone. Osteoblasts are activated and synthesize new lamellar bone concentrically, gradually closing the diameter of the tunnel (Figure 1.8A). After about four weeks, bone production stops as the tunnel is closed, leaving behind a vascularized cavity called an osteon that runs parallel to the long axis [96]. It is believed that the greater number of osteons that cross the site of injury, the greater the ultimate strength [79].

1.5.1.1.3   PNS Remodeling

In the PNS, once regenerating axons find their target, the tissue matures as axons gradually increase in thickness through neurofilament synthesis (Figure 1.7C). The axonal diameter and myelin sheath thickness of regenerated neurons are usually thinner and never reach normal preinjury levels [39]. Daughter axons that do not make contact with the target are cleaved off. During regeneration, a parent axon sprouts an average of 3 daughter axons, although up to 25 daughter axons have been observed [39]. Greater amounts of neural death generally lead to greater sprouting since there is less competition for access to the target [39]. The PNS is capable of maintaining a regenerative response at least 12 months after injury [124]. Schwann cell scaffolds that remain uninnervated slowly shrink in diameter, and if they do not receive a regenerating axon, they lose supporting ability as they are progressively filled with fibrous tissue [124].

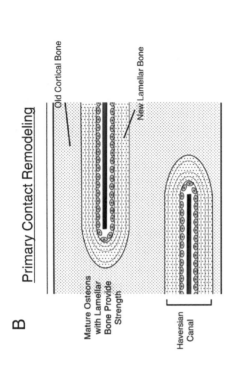

B Primary Contact Remodeling

Old Cortical Bone

New Lamellar Bone

Mature Osteons
with Lamellar
Bone Provide
Strength

Haversian
Canal

A Primary Gap Remodeling

Mature Osteon

Cutting Cone
Moves across
Scaffolding

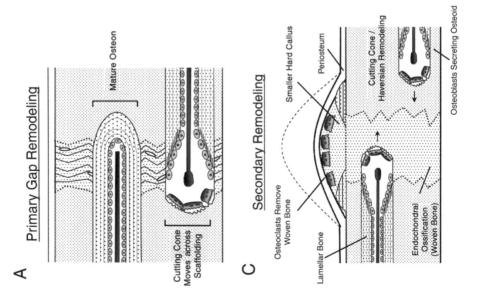

C Secondary Remodeling

Smaller Hard Callus

Periosteum

Osteoclasts Remove
Woven Bone

Lamellar Bone

Cutting Cone /
Haversian Remodeling

Endochondral
Ossification
(Woven Bone)

Osteoblasts Secreting Osteoid

FIGURE 1.8 Primary versus secondary bone remodeling. (A) Lamellar bone deposited perpendicular to the bone's long axis during primary repair is used as a scaffold for cutting cones. Osteoclasts create tunnels through which new blood vessels follow, stimulating osteoblasts to produce new osteoid that becomes mineralized to form lamellar bone. A mature osteon forms when osteoclast and osteoblast activity becomes quiescent. (B) In primary contact remodeling the cutting cones initiated during repair mature, forming "mature osteons." The larger the number of osteons, the greater the tissue strength of the repaired tissue. (C) In secondary remodeling, the mineralized hard callus near the periosteum not needed for structural support is removed and remodeled by osteoclasts arriving from the periosteum. Within the cortex, cutting cones are formed as described in primary remodeling. Less organized woven bone within the wound gap is replaced with oriented lamellar bone. The ultimate endpoint of both primary and secondary bone healing is tissue with completely restored function.

Axonal repair and remodeling depends on the severity of the trauma. Most injuries in the PNS are stretch related, where the tensile force applied to the tissue exceeds the nerve's ability to stretch. Generally in these injuries, the outer connective tissue layer of the nerve, the endoneurium, is intact, and the axon is able to regenerate as illustrated previously (Figure 1.3C) [124]. However, in more traumatic injuries such as lacerations, the endoneurium and the axon are severed and the tissue usually fails to regenerate. After traumatic injuries the fibroblasts contained within the endoneurium proliferate and produce a collagenous scar around the nerve trunk during the repair and remodeling phases. This collagenous scar misdirects or blocks axonal regeneration [124]. Sprouting axons that cannot find their target or contact with a Schwann cell conduit are either cleaved or grow into a disorganized mass, resulting in the formation of a neuroma. Meanwhile, the target muscle remains inactive, and neural cell bodies atrophy and eventually die. Even when the endoneurium is not severed, regeneration of the target site fails more often than not, and even in the best case, regeneration of a peripheral nerve does not fully restore the tissue back to its original status since there is inevitably inaccuracy during the attempted return of an axon to its original target [88].

Surgical insertion of an autologous nerve graft can be used to repair PNS lesions that are too large to be bridged by Schwann cells. The purpose of the graft is to reconnect damaged nerves end-to-end without causing tension. Suturing is used to connect the connective tissue sheath of the damaged nerve to the autologous graft. Freshly injured tissue sheaths do not hold sutures very well, so surgical repair is generally not performed until three weeks after injury, when the sheaths have had time to thicken. The major drawback of this method of repair is that harvest of the autologous nerve entails sacrificing one or more nerves. A number of groups are working to engineer synthetic nerve grafts for PNS repair [86,125].

1.5.1.2 Secondary Remodeling

1.5.1.2.1 Full-Thickness Cutaneous Tissue Remodeling

The major goal of secondary wound remodeling is to reduce the amount of excess ECM and align the ECM through contraction. If the extent of the wound is relatively small and the reorganization of the ECM is efficient, relatively little contracture occurs, and little or no scarring is observed. However, in larger injuries where there is more extensive tissue repair through ECM production, significant remodeling is required, which results in scarring. Remodeling occurs over a long time period. This phase can overlap with the repair phase, as it can begin as early as 1 week after injury and can last as long as 2 years, depending on the wound.

During repair, fibroblasts migrate into the site of injury and produce ECM to replace lost tissue. The tractional forces that fibroblasts create as they move through the wounded tissue generate mechanical tension, which promotes wound closure [126]. The remodeling phase officially begins when TGF-β and other cytokines released from platelets and activated macrophages cause fibroblasts to differentiate into a more contractile phenotype called myofibroblasts. Myofibroblasts are characterized by their expression of alpha smooth muscle actin and production of collagen I [28,127]. Once activated, they increase their cytoskeletal stress fibers and focal adhesions, providing constant tension to contract the wound bed (for review see [127]). Myofibroblast contraction is similar to that of smooth muscle cells, though the mode of activation is vastly different. Smooth muscle cells contract because of elevations of Ca^{2+}. Since Ca^{2+} levels around cells can change rapidly, the contractility of smooth-muscle cells can change quickly [128]. Myofibroblasts are believed to be regulated by the Rho–Rho kinase pathway, which is less transient, and causes a longer-lasting contraction force. Contractures of 10 to 20 μm per day are possible without the need to generate large amounts of force because of the low basal tension level of skin [127]. Myofibroblasts continue to support loads in contracted tissue until ECM is produced and crosslinked, leading to stress shielding [127]. Collagen fibers gradually thicken and, along with myofibroblasts, they become oriented parallel to the wound bed along lines of stress, resulting in the appearance of striated scar tissue (Figure 1.7D). This is in direct contrast to the basket weave pattern seen in uninjured skin [129].

Once wound contraction occurs, stress relaxation causes myofibroblasts to return to a quiescent state. The cells then receive a signal to undergo apoptosis, transforming the wound from cell-rich granulation tissue to cell-poor scar tissue with an excess of ECM [130]. At the same time, the capillary density gradually diminishes and the wound loses its pink color, becoming progressively paler [27]. The ultimate end point of the remodeling process is the formation of acellular scar tissue that is poorly reorganized into dense parallel bundles, as opposed to the tightly woven meshwork of normal dermal tissue (Figure 1.7D) [72]. Dermal structures such as hair follicles, sweat glands, and sebaceous glands that are lost during injury are not regenerated [69]. After three months unmineralized tissue can have a maximum of 80% of the strength of unwounded tissue [68]. This is generally the highest level of strength that a healed tissue can achieve.

1.5.1.2.2   Unstabilized Bone Remodeling

The ultimate goal of secondary remodeling of mineralized tissue is full structural restoration with little to no scarring (Figure 1.5E). While remodeling can last up to 6 months, the endpoint is healed tissue that is remarkably similar to noninjured tissue in terms of robustness but may appear slightly less organized [78]. The only major difference between primary and secondary mineralized tissue remodeling is the extensive bone removal required to remove the excess callus produced during bone repair.

In secondary healing, osteoclasts arrive at the site in need of remodeling via the circulation. They recognize and attach to cell adhesion proteins such as osteopontin, osteocalcin, and osteonectin [83]. Osteoclasts are present not only to remove excess bone not needed for structural support, but also to digest woven bone synthesized from the soft callus during repair so it can be replaced with lamellar bone aligned in response to stress. Like primary remodeling, secondary remodeling occurs mainly through cutting cones. Within the cortex, woven and necrotic lamella bone is removed by osteoclasts. Osteoblasts follow closely and produce new compact lamellar bone, giving the tissue greater strength (Figure 1.8C). Each concentric 100-μm-thick lamellar layer is directed in a specific manner to buttress fracture fragments [96]. Secondary remodeling is altered from primary remodeling because of the large external callus formed during repair. The external callus is formed outside the cortex and is generally remodeled by osteoclasts without the formation of cutting cones and osteons. This is possible because osteoclasts have immediate access to woven bone of the external callus via the periosteum (Figure 1.8C).

Much of the bone removed during remodeling of the cortex is replaced by lamellar bone. During remodeling, mechanical loads applied to mineralized tissue are capable of generating signals at the cellular level to increase bone production (for review see [80]). As the most abundant cell in bone, osteocytes are enclosed within the bone matrix and are capable of communicating with neighboring cells through their network of processes connected by gap junctions. Mechanical stress within mineralized tissue causes fluid shear on osteocytes and causes an influx of extracellular calcium ions as well as ATP release leading to ion channel activation [80]. This in turn may stimulate lamellar bone synthesis by activating specific prostaglandins and increase nitric oxide production [80]. Osteoblast activity has been shown to increase in the presence of certain prostaglandins [131], while nitric oxide is known to inhibit bone resorption by osteoclasts [132].

1.5.2 CNS REMODELING

Remodeling in the CNS is limited. Because of the need to protect the CNS from the body's robust inflammatory responses, reactive astrocytic processes become further intertwined, forming a dense sheath around the wound site (Figure 1.7A). During the repair stages, GFAP reactivity is seen approximately 400 µm outside the lesion. During the remodeling stages, GFAP reactivity is only observed less than 100 µm from the site of tissue damage [133]. Glial scar density increases as hypertrophic astrocytes become more condensed around the site of damage (Figure 1.9) [11].

Inside the lesion, removal of the fibrin clot and the necrotic neurons and supporting glial cells is completed by microglia and macrophages (Figure 1.7A) [105].

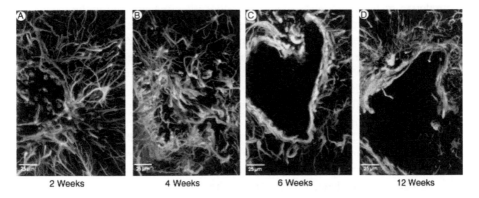

| 2 Weeks | 4 Weeks | 6 Weeks | 12 Weeks |

FIGURE 1.9 Time course of glial scar formation at four time points as imaged by GFAP staining. At 2- and 4-week time points, the astrocytic processes fall back into the void left by the probe extraction before tissue processing. By 6 weeks, the processes have interwoven to form a stronger, more dense sheath surrounding the implant. Minimal changes between the 6- and 12-week time points indicate the glial scar completion within 6 weeks. (Reprinted from *Exp. Neurol.*, 156, 33–49, 1999. With permission from Elsevier.)

Once complete, microglia and macrophages undergo apoptosis, leaving a cerebral spinal fluid–filled cyst in the center of the lesion where the initial wound occurred [105]. The cyst is bordered by a thin, dense layer of reactive astrocytes that serve as a barrier between healthy and lost tissue and may help protect neurons outside the injury. The axons of neurons protected outside the glial scar cannot regrow into lost tissue because of inhibitory molecules within the scar, and thus tissue function is never regained (Figure 1.7A).

1.6   CONCLUSION

The cellular reaction after injury depends on the tissue type as well as the extent of the wound. In injury to CNS tissue that damages neurons and the supporting glial cells, the body's response is unforgiving, as regeneration of lost neurons is not possible. Activated astrocytes wall off the lesion, creating a glial scar. These activated astrocytes may prevent further tissue damage, although neuron axonal regrowth is inhibited. In contrast, in non-CNS tissue, a single tissue type can have multiple responses depending on the magnitude of injury. For example, a superficial skin wound often has lower levels of inflammatory infiltrate and can undergo primary healing, while a deeper wound with more extensive tissue damage and cellular loss will undergo secondary healing. This chapter focused on the primary healing response seen after injuries in skin, bone, and the PNS where the general tissue architecture is maintained after injury, allowing the tissues to undergo repair predominantly by the regeneration of lost cells to restore the tissue's normal structure. In secondary healing, the wounds are often deeper and the general tissue structure is compromised. The inflammatory response may be more intense and prolonged as well, recruiting fibroblasts and endothelial cells that produce granulation tissue. Within granulation tissue, the deposition of collagen by fibroblasts changes the structure and function of the tissue and eventually leads to scar formation upon the completion of the remodeling stage. Secondary healing in bone is the exception, as bone tissue is capable of removing the initial fibrous callus and replacing it with lamellar bone. Other examples of scarring after secondary healing occur in tissues that are not capable of undergoing regeneration such as the heart.

In summary, while the overall outcome of CNS and non-CNS tissue wound healing has been described, for therapeutic purposes in the CNS, it is important to consider the different junctures within the wound healing process at which the response can be modified to push the body's wound healing response toward a desired outcome. These points of intervention are mainly within the stages hemostasis, inflammation, and repair. Within each stage there are points that may be useful for modulating the wound healing response. For instance, to limit extensive clot formation, it has been suggested that there are three main stages within clot formation for which therapeutics can be developed: initiation (platelet activation), propagation (coagulation cascade), and fibrin formation (thrombin) [134]. During inflammation, modulation strategies center around limiting immune cell activation and reducing inflammatory cell migration into the wound [67]. Since neurons within the CNS are especially sensitive to inflammatory damage [48], and because chronic inflammation is involved in many other diseases outside the CNS, there has been a great deal of effort to address mechanisms through which inflammation is dampened, with the hope of achieving a balance between infection prevention and resolution of inflammation [67]. Finally, to alter the CNS repair process for improving CNS function after injury, researchers have attempted to reduce astrocyte activity [135] and stimulate neural regeneration [111].

CNS wound healing interventions focusing on altering the glial scar and the inflammatory processes should be approached with caution. A decrease in glial scaring could result in unnecessary tissue damage due to the inability to reestablish normal homeostasis and repair of the blood–brain barrier [109,136]. One of the major concerns regarding antiinflammatory approaches is loss of defense against infection, although methods so far have not shown substantial increases in the susceptibility to infection, suggesting that it is difficult to completely shut off the inflammatory process because of redundancies in the pathway [67].

REFERENCES

1. Wisniewski, N. and Reichert, M. Methods for reducing biosensor membrane biofouling. *Colloids and Surfaces B: Biointerfaces* 18(3-4), 197, 2000.
2. Gerritsen, M. and Jansen, J. Performance of subcutaneously implanted glucose sensors: a review. *Journal of Investigative Surgery* 11, 163, 1998.
3. Wilson, G. S. and Gifford, R. Biosensors for real-time in vivo measurements. *Biosensors and Bioelectronics* 20(12), 2388, 2005.
4. Updike, S., Shults, M., and Rhodes, R. Principles of long-term fully implanted sensors with emphasis on radiotelemetric monitoring of blood glucose from inside a subcutaneous foreign body capsule (FBC). In Fraser, D. (ed), *Biosensors in the Body*. New York, Wiley, 1997, p. 117.
5. Reichert, W., and Sharkawy, A. Biosensors. In von Recum, A. (ed), *Handbook of Biomaterials Evaluation*. Ann Arbor, MI, Taylor & Francis, 1999.
6. Wisniewski, N., Moussy, F., and Reichert, W. M. Characterization of implantable biosensor membrane biofouling. *Fresenius' Journal of Analytical Chemistry* 366(6-7), 611, 2000.
7. Polikov, V. S., Tresco, P. A., and Reichert, W. M. Response of brain tissue to chronically implanted neural electrodes. *Journal of Neuroscience Methods* 148(1), 1, 2005.
8. Biran, R., Martin, D. C., and Tresco, P. A. Neuronal cell loss accompanies the brain tissue response to chronically implanted silicon microelectrode arrays. *Experimental Neurology* 195(1), 115, 2005.
9. Kim, Y.-T., Hitchcock, R. W., Bridge, M. J., and Tresco, P.A. Chronic response of adult rat brain tissue to implants anchored to the skull. *Biomaterials* 25(12), 2229, 2004.
10. Szarowski, D. H., Andersen, M. D., Retterer, S., Spence, A. J., Isaacson, M., Craighead, H. G., Turner, J. N., and Shain, W. Brain responses to micro-machined silicon devices. *Brain Research* 983(1-2), 23, 2003.
11. Turner, J. N., Shain, W., Szarowski, D. H., Andersen, M., Martins, S., Isaacson, M., and Craighead, H. Cerebral astrocyte response to micromachined silicon implants. *Experimental Neurology* 156(1), 33, 1999.
12. Carmena, J. M., Lebedev, M. A., Crist, R. E., O'Doherty, J. E., Santucci, D. M., Dimitrov, D. F., Patil, P. G., Henriquez, C. S., and Nicolelis, M. A. L. Learning to control a brain–machine interface for reaching and grasping by primates. *PLoS Biology* 1(2), 193, 2003.

13. Hochberg, L. R., Serruya, M. D., Friehs, G. M., Mukand, J. A., Saleh, M., Caplan, A. H., Branner, A., Chen, D., Penn, R. D., and Donoghue, J. P. Neuronal ensemble control of prosthetic devices by a human with tetraplegia. *Nature* 442(7099), 164, 2006.

14. Clark, R., Schiding, J., Kaczorowski, S., Marion, D., and Kochanek, P. Neurtrophil accumulation after traumatic brain injury in rats: comparison of weight drop and controlled cortical impact models. *Journal of Neurotrauma* 11, 499, 1994.

15. Purves, D., Augustine, G., Fitzpatrick, D., Katz, L., LaMantia, A., McNamara, J., and Williams, W. *The Organization of the Nervous System.* Sunderland, MA, Sinauer Associates, 2001.

16. Rubin, L. L. and Staddon, J. M. The cell biology of the blood-brain barrier. *Annual Review of Neuroscience* 22(1), 11, 1999.

17. Schwartz, M. Macrophages and microglia in central nervous system injury: are they helpful or harmful? *Journal of Cerebral Blood Flow and Metabolism* 23, 385, 2003.

18. Seung, U. K. de Vellis, J. Microglia in health and disease. *Journal of Neuroscience Research* 81(3), 302, 2005.

19. Ruggeri, Z. M. Platelets in atherothrombosis. *Nature Medicine* 8(11), 1227, 2002.

20. Standeven, K. F., Ariens, R. A. S., and Grant, P. J. The molecular physiology and pathology of fibrin structure/function. *Blood Reviews* 19(5), 275, 2005.

21. del Zoppo, F., Poeck, K., Pessin, M., Wolpert, S., Furlan, A., Febert, A., Alberts, M., Zivin, J., Wechsler, L. and, Busse, O. Recombinant tissue plasminogen activator in acute thrombotic and embolic stroke. *Annals of Neurology* 68, 642, 1992.

22. Rosenberg, R. D., and Aird, W. C. Vascular-bed-specific hemostasis and hypercoagulable states. *New England Journal of Medicine* 340(20), 1555, 1999.

23. Heldin, C. H. Westermark, B., and Wasteson. A. Platelet-derived growth factor. Isolation by a large-scale procedure and analysis of subunit composition. *Biochemistry Journal* 193(3), 907, 1981.

24. Assoian, R. K., Komoriya, A., Meyers, C. A., Miller, D. M., and Sporn, M. B. Transforming growth factor-beta in human platelets. Identification of a major storage site, purification, and characterization. *Journal of Biological ChemIstry* 258(11), 7155, 1983.

25. Mohle, R., Green, D., Moore Malcolm, A. S., Nachman, R. L., and Rafii, S. Constitutive production and thrombin-induced release of vascular endothelial growth factor by human megakaryocytes and platelets. *Proceedings of the National Academy of Sciences USA* 94(2), 663, 1997.

26. Anderson, J. Biological responses to materials. *Annual Review Materials Research* 31, 81, 2001.

27. Fine, N., and Mustoe, T. *Wound Healing.* Philadelphia, Lippincott Williams & Wilkins, 2006.

28. Martin, P. Wound healing—aiming for perfect skin regeneration. *Science* 276(5309), 75, 1997.

29. Yager, D., and Nwomeh, B. The proteolytic environment of chronic wounds. *Wound Repair Regeneration* 7, 433, 1999.

30. Diacovo, T. G., Roth, S. J., Buccola, J. M., Bainton, D. F. and, Springer, T. A. Neutrophil rolling, arrest, and transmigration across activated, surface-adherent platelets via sequential action of P-selectin and the beta 2-integrin CD11b/CD18. *Blood* 88(1), 146, 1996.

31. Springer, T. A. Traffic signals for lymphocyte recirculation and leukocyte emigration: the multistep paradigm. *Cell* 76(2), 301, 1994.

32. Martin, P., and Leibovich, S. J. Inflammatory cells during wound repair: the good, the bad and the ugly. *Trends in Cell Biology* 15(11), 599, 2005.

33. Shukaliak, J., and Dorovini-Zis, K. Expression of the [beta]-chemokines RANTES and MIP-1[beta] by human brain microvessel endothelial cells in primary culture. Journal Neuropathology *Experimental Neurology* 59(5), 339, 2000.

34. Simpson, D. M., and Ross, R. The neutrophilic leukocyte in wound repair: a study with antineutrophil serum. *Journal Clinical Investigation* 51(8), 2009, 1972.

35. Leibovich, S. J., and Ross, R. The role of the macrophage in wound repair. A study with hydrocortisone and antimacrophage serum. *Americal Journal of Pathology* 78(1), 71, 1975.

36. Hickey, W. Basic principles of immunological surveillance of the normal central nervous system. *Glia* 36(2), 118, 2001.

37. Perry, V. H., Brown, M. C., and Gordon, S. The macrophage response to central and peripheral nerve injury. A possible role for macrophages in regeneration. *Journal of Experimental MedIcine* 165(4), 1218, 1987.

38. Avellino, A., Hart, D., Dailey, A., MacKinnon, M., Ellegala, D., and Kliot, M. Differential macrophage response in the peripheral and central nervous system during Wallerian degeneration of Axons. *Experimental Neurology* 136, 183, 1995.

39. Bisby, M. Regeneration of peripheral nervous system axons. In Waxman, S., Kocsis, J., and Stys, P. (eds), *The Axon: Structure, Function, and Pathophysiology*. New York, Oxford University Press, 1995, p. 553.

40. Fawcett, J. W. The glial response to injury and its role in the inhibition of CNS repair. In Bahr, M. (ed), *Brain Repair*, Vol. 557. New York, Springer Science, 2006, p. 11.

41. Berry. M., Butt, A., and Logan, A. Cellular responses to penetrating CNS injury. In Berry, M., and Logan, A. (eds), *CNS Injuries: Cellular Responses and Pharmacological Strategies*. Boca Raton, FL, CRC Press, 1999, p. 1.

42. Griffin, J., George, E., Hsieh, S., and Glass, J. Axonal degeneration and disorders of the axonal cytoskeleton. In Waxman, S., Kocsis, J., and Stys, P. (eds), *The Axon: Structure, Function, and Pathophysiology*. New York, Oxford University Press, 1995, p. 375.

43. Ford, A. L., Goodsall, A. L., Hickey, W. F., and Sedgwick, J. D. Normal adult ramified microglia separated from other central nervous system macrophages by flow cytometric sorting. Phenotypic differences defined and direct ex vivo antigen presentation to myelin basic protein- reactive CD4+ T cells compared. *Journal of Immunology* 154(9), 4309, 1995.

44. Streit, W. Microglial cells. In Kettenmann, H. and Ransom, B. (eds), *Neuroglia*. Oxford, U.K., Oxford University Press, 2005.

45. Olson, J. K. and Miller, S. D. Microglia initiate central nervous system innate and adaptive immune responses through multiple TLRs. *Journal of Immunology* 173(6), 3916, 2004.

46. Hayes, G., Woodroofe, M., and Cuzner, M. Microglia are the major cell type expressing MHC class II in human white matter. *Journal of the Neurological Sciences* 80, 25, 1987.

47. Fujita, T., Yoshimine, T., Maruno, M., and Hayakawa, T. Cellular dynamics of macrophages and microglial cells in reaction to stab wounds in rat cerebral cortex. *Acta Neurochirurgica* 140(3), 275, 1998.

48. Fitch, M. T., Doller, C., Combs, C. K., Landreth, G. E., and Silver, J. Cellular and molecular mechanisms of glial scarring and progressive cavitation: in vivo and in vitro analysis of inflammation-induced secondary injury after CNS trauma. *Journal of Neuroscience* 19(19), 8182, 1999.

49. Babcock, A. A., Kuziel, W. A., Rivest, S., and Owens, T. Chemokine expression by glial cells directs leukocytes to sites of axonal injury in the CNS. *Journal of Neuroscience* 23(21), 7922, 2003.

50. Giulian, D., Li, J., Li, X., George, J., and Rutecki, P. The impact of microglia-derived cytokines upon gliosis in the CNS. *Developmental Neuroscience* 16(3-4), 128, 1994.

51. Giulian, D., Li, J., Leara, B., and Keenen, C. Phagocytic microglia release cytokines and cytotoxins that regulate the survival of astrocytes and neurons in culture. *Neurochemistry International* 25(3), 227, 1994.

52. Woodroofe, M., Sarna, G., Wadhwa, M., Hayes, G., Loughlin, A., Tinker, A., Cuzner, M. Detection of interleukin-1 and interleukin-6 in adult-rat brain, following mechanical injury, by in vivo microdialysis—evidence of a role for microglia in cytokine production. *Journal Neuroimmunology* 33(3), 227, 1991.

53. Zielasek, J., Muller, B., and Hartung, H.-P. Inhibition of cytokine-inducible nitric oxide synthase in rat microglia and murine macrophages by methyl-2,5-dihydroxycinnamate. *Neurochemistry International* 29(1), 83, 1996.

54. Giulian, D., Baker, T. J. Characterization of ameboid microglia isolated from developing mammalian brain. *Journal Neuroscience* 6(8), 2163, 1986.

55. Morgan, S. C., Taylor, D. L., and Pocock, J. M. Microglia release activators of neuronal proliferation mediated by activation of mitogen-activated protein kinase, phosphatidylinositol-3-kinase/Akt and delta-Notch signalling cascades. *Journal of Neurochemistry* 90(1), 89, 2004.

56. Polazzi, E., Gianni, T., and Contestabile, A. Microglial cells protect cerebellar granule neurons from apoptosis: evidence for reciprocal signaling. *Glia* 36(3), 271, 2001.

57. Elkabes, S., DiCicco-Bloom, E. M., and Black, I. B. Brain microglia/macrophages express neurotrophins that selectively regulate microglial proliferation and function. *Journal of Neuroscience* 16(8), 2508, 1996.

58. Nakajima, K., Honda, S., Tohyama, Y., Imai, Y., Kohsaka, S., and Kurihara, T. Neurotrophin secretion from cultured microglia. *Journal of Neuroscience Research* 65(4), 322, 2001.

59. Mucke, L., Oldstone, M., Morris, J., and Nerenberg, M. Rapid activation of astrocyte-specific expression of GFAP-lacZ transgene by focal injury. *New Biology* 3(5), 465, 1991.

60. Norenberg, M. The reactive astrocyte. In Aschner, M. and Costa, L. (eds). *The Role of Glia in Neurotoxicity*. Boca Raton, FL, CRC Press, 2005, p. 73.

61. George, R., and Griffin, J. W. Delayed macrophage responses and myelin clearance during Wallerian degeneration in the central nervous system: the dorsal radiculotomy model. *Experimental Neurology* 129(2), 225, 1994.

62. Niklason, L. E., Abbott, W., Gao, J., Klagges, B., Hirschi, K. K., Ulubayram, K., Conroy, N., Jones, R., Vasanawala, A., Sanzgiri, S., and Langer, R. Morphologic and mechanical characteristics of engineered bovine arteries. *Journal of Vascular Surgery* 33(3), 628, 2001.

63. Wolburg, H., Wolburg-Buchholz, K., and Engelhardt, B. Diapedesis of mononuclear cells across cerebral venules during experimental autoimmune encephalomyelitis leaves tight junctions intact. *Acta Neuropathologica* 109(2), 181, 2005.

64. Butcher, E., Williams, M., Youngman, K., Rott, L., and Briskin, M. Lymphocyte trafficking and regional immunity. *Advances in Immunology* 72, 209, 1999.

65. Bauer, J., Bradl, M., Hickey, W. F., Forss-Petter, S., Breitschopf, H., Linington, C., Wekerle, H., and Lassmann, H. T-cell apoptosis in inflammatory brain lesions: destruction of T cells does not depend on antigen recognition. *American Journal of Pathology* 153(3), 715, 1998.

66. Loddick, S., and Rothwell, N. Cytokines and neurodegeneration. In Rothwell, N. and Loddick, S. (eds), *Immune and Inflammatory Responses in the Nervous System.* Oxford, U.K., Oxford University Press, 2002, p. 90.
67. Henson, P. M. Dampening inflammation. *Nature Immunology* 6(12), 1179, 2005.
68. Witte, M. and Abarbul, A. General principles of wound healing. *Surgical Clinics of North America* 3, 509, 1997.
69. Mehendale, F. and Martin, P. The cellular and molecular events of wound healing. In Falanga, V. (ed). *Cutaneous Wound Healing.* London, Martin Dunitz, 2001.
70. Saarialho-Kere, U. K., Chang, E. S., Welgus, H. G., and Parks, W. C. Distinct localization of collagenase and tissue inhibitor of metalloproteinases expression in wound healing associated with ulcerative pyogenic granuloma. *Journal of Clinical Investigations* 90(5), 1952–1957, 1992.
71. Mogford, J., and Mustoe, T. Experimental models of wound healing. In Falanga, V. (ed). *Cutaneous Wound Healing.* London, Martin Dunitz, 2001, p. 109.
72. Xu, J., and Clark, R. Integrin regulation in wound repair. In Garg, H. and Longaker, M. (eds), *Scarless Wound Healing.* New York, Marcel Dekker, 2000.
73. Romer. J., Bugge, T., Pyke, C., Lund, L., Flick, M., Degen, J., and Dano, K. Impaired wound healing in mice with a disrupted plasminogen gene. Nature Medicine 2(3), 287, 1996.
74. Grinnell, F. Wound repair, keratinocyte activation and integrin modulation. *Journal of Cell Science* 101(1), 1, 1992.
75. Matoltsy, A. and Viziam, C. Further observations on epithelialization of small wounds: an autoradiographic study of incorporation and distribution of 3H-thymidine in the epithelium covering skin wounds. *Journal of Investigative Dermatology* 55(1), 20, 1970.
76. Sfeir, C., Ho, L., Doll, B., Azari, K., and Hollinger, J. Fracture repair. In Lieberman, J. and Friedlaender, G. (eds). *Bone Regeneration and Repair: Biology and Clinical Applications.* Totowa, NJ, Humana Press, 2005, p. 21.
77. Delmas, P. and Malaval, L. The proteins of bone. In Mundy, G. (ed). *Physiology and Pharmacology of Bone.* New York, Springer-Verlag, 1993, p. 673.
78. Mast, B. Wound healing in other tissues. *Surgical Clinics of North America* 77(3), 529, 1997.
79. Chao, E., Aro, H., Lewallen, D., and Kelly, P. The effect of rigidity on fracture healing in external fixation. *Clinical Orthopaedics and Related Research* 241, 24, 1989.
80. Robling, A., Castillo, A., and Turner, C. Biomechanical and molecular regulation of bone remodeling. *Annual Review of Biomedical Engineering* 8, 455, 2006.
81. McHugh, K. P., Hodivala-Dilke, K., Zheng, M.-H., Namba, N., Lam, J., Novack, D., Feng, X., Ross, F. P., Hynes, R. O., and Teitelbaum, S. L. Mice lacking {beta}3 integrins are osteosclerotic because of dysfunctional osteoclasts. *Journal of Clinical Investigation* 105(4), 433, 2000.
82. Teitelbaum, S. L. Bone resorption by osteoclasts. *Science* 289(5484), 1504, 2000.
83. Lee, C., and Einhorn, T. The bone organ system: form and function. In Marcus, R. (ed), *Osteoporosis*, Vol. 1. San Diego, CA: Academic Press, 2001.
84. McKibbin, B. The biology of fracture healing in long bones. *Journal of Bone and Joint Surgery* 60B, 150, 1978.
85. Lee, S., and Wolfe, S. Peripheral nerve injury and repair. *Journal of the American Academy of Orthopaedic Surgery* 8(4), 243, 2000.
86. Schmidt, C. E. and Leach, J. B. Neural tissue engineering: strategies for repair and regeneration. *Annual Review of Biomedical Engineering* 5(1), 293, 2003.

87. Sunderland, S. *Nerve Injuries and Their Repair: A Critical Appraisal.* New York, Churchill Livingstone, 1991.
88. Frostick, S., Yin, Q., and Kemp, G. Schwann cells, neurotrophic factors, and peripheral nerve regeneration. *Microsurgery* 18, 397, 1998.
89. Dodd, J. and Jessell, T. Axon guidance and the patterning of neuronal projections in vertebrates. *Science* 242, 692, 1988.
90. Gundersen, R. and Barrett, J. Characterization of the turning response of dorsal root neurites toward nerve growth factor. *Journal of Cell Biology* 87, 546, 1980.
91. Grabham, P. W., Foley, M., Umeojiako, A., and Goldberg, D. J. Nerve growth factor stimulates coupling of beta1 integrin to distinct transport mechanisms in the filopodia of growth cones. *Journal of Cell Science* 113(17), 3003, 2000.
92. Grinnell, F., Billingham, R., and Burgess, L. Distribution of fibronectin during wound healing in vivo. *Journal of Investigative Dermatology* 76(3), 181, 1981.
93. Eckes, B., Aumailley, M., and Kreig, T. Collagens and the reestablishment of dermal integrity. In Clark, R. (ed), *The Molecular and Cellular Biology of Wound Repair.* New York, Plenum Press, 1996, p. 493.
94. Singer, A. J. and Clark, R. A. F. *Cutaneous wound healing. New England Journal of Medicine* 341(10), 738, 1999.
95. Chen, W. Y. J. and Abatangelo, G. Functions of hyaluronan in wound repair. *Wound Repair and Regeneration* 7(2), 79, 1999.
96. Doll, B., Sfeir, C., Azari, K., Holland, S., and Hollinger, J. Craniofacial repair. In Lieberman, J., Friedlaender, G. (eds), *Bone Regeneration and Repair: Biology and Clinical Applications.* Totowa, NJ, Humana Press, 2005, p. 337.
97. Nakahara, H., Bruder, S. P., Haynesworth, S. E., Holecek, J. J., Baber, M. A., Goldberg, V. M., and Caplan, A. I. Bone and cartilage formation in diffusion chambers by subcultured cells derived from the periosteum. *Bone* 11(3), 181, 1990.
98. Wornom, I. and Buchman, S. Bone and cartilaginous tissue. In Cohen, K., Diegelmann, R., and Lindglad, W. (eds), *Wound Healing: Biochemical and Clinical Aspects.* Philadelphia, WB Saunders, 1992, p. 356.
99. Kon, T., Cho, T., Aizawa, T., Yamazaki, M., Nooh, N., Graves, D., Gerstenfeld, L., and Einhorn, T. Expression of osteoprotegerin, receptor activator of NF-kappaB ligand (osteoprotegerin ligand) and related proinflammatory cytokines during fracture healing. *Journal of Bone Mineral Research* 16(6), 1004, 2001.
100. Buckwalter, J., Einhorn, T., and Marsh, J. Bone and joint healing. In Bucholz, R. and Heckman, J. (eds), *Rockwood and Green's Fractures in Adults*, Vol. 1. Philadelphia, Lippincott Williams and Wilkins, 2001.
101. Poynton, A. and Lane, J. Bone healing and failure. In Petite, H., Quarto, R. (eds), *Engineered Bone.* Georgetown, TX, Landes Bioscience, 2005.
102. Miclau, T., Schneider, R., Eames, B., and Helms, J. Common molecular mechanisms regulating fetal bone formation and adult fracture repair. In Lieberman, J. and Friedlaender, G. (eds), *Bone Regeneration and Repair: Biology and Clinical Applications.* Totowa, NJ, Humana Press, 2005, p. 45.
103. Zhao, B.-Q., Wang, S., Kim, H.-Y., Storrie, H., Rosen, B. R., Mooney, D. J., Wang, X., and Lo, E. H. Role of matrix metalloproteinases in delayed cortical responses after stroke. *Nature Medicine* 12(4), 441, 2006.
104. Silver, J. and Miller, J. H. Regeneration beyond the glial scar. *Nature Reviews Neuroscience* 5(2), 146, 2004.
105. Clark, R., Lee, E., Fish, C., White, R., Price, W., Jonak, Z., Feuerstein, G., and Barone, F. Development of tissue damage, inflammation and resolution following stroke: an immunohistochemical and quantitative planimetric study. *Brain Research Bulletin* 31, 565, 1993.

106. Frosch, M., Anthony, D., and DeGirolami, V., The central nervous system. In Kumar, V., Abbas, A., and Fausto, N. (eds.), *Robbins and Cotran Pathologic Basis of Disease*. Philadelphia, Elsevier, 2005.
107. Fawcett, J. W. and Asher, R. A. The glial scar and central nervous system repair. *Brain Research Bulletin* 49(6), 377, 1999.
108. Berry, M. and Logan, A. Transforming growth factor-beta and CNS scarring. In Berry, M. and Logan, A. (eds.), *CNS Injuries: Celluar Responses and Pharmacological Strategies*. Boca Raton, FL, CRC Press, 1999.
109. Faulkner, J. R., Herrmann, J. E., Woo, M. J., Tansey, K. E., Doan, N. B., and Sofroniew, M. V. Reactive astrocytes protect tissue and preserve function after spinal cord injury. *Journal of Neuroscience* 24(9), 2143, 2004.
110. Zurn, A. and Bandtlow, C. Regeneration failure in the CNS: cellular and molecular mechanisms. In Bahr, M. (ed), *Brain Repair*, Vol. 557. New York, Springer Science, 2006.
111. Fawcett, J. W. Overcoming inhibition in the damaged spinal cord. *Journal of Neurotrauma* 23(3-4), 371, 2006.
112. McKeon, R. J., Jurynec, M. J., and Buck, C. R. The chondroitin sulfate proteoglycans neurocan and phosphacan are expressed by reactive astrocytes in the chronic CNS glial scar. *Journal of Neuroscience* 19(24), 10778, 1999.
113. Davies, S. J. A., Goucher, D. R., Doller, C. and Silver, J. Robust regeneration of adult sensory axons in degenerating white matter of the adult rat spinal cord. *Journal of Neuroscience* 19(14), 5810, 1999.
114. Davies, Y., Lewis, D., Fullwood, N. J., Nieduszynski, I. A., Marcyniuk, B., Albon, J., and Tullo, A. Proteoglycans on normal and migrating human corneal endothelium. *Experimental Eye Research* 68(3), 303, 1999.
115. Sandvig, A., Berry, M., Barrett, L., Butt, A., and Logan, A. Myelin-, reactive glia-, and scar-derived CNS axon growth inhibitors: expression, receptor signaling, and correlation with axon regeneration. *Glia* 46(3), 225, 2004.
116. David, S. and Aguayo, A. Axonal elongation into PN "bridges" after central nervous system injury in adult rats. *Science* 214, 931, 1981.
117. Davies, S. J. A., Field, P. M. and Raisman, G. Regeneration of cut adult axons fails even in the presence of continuous aligned glial pathways. *Experimental Neurology* 142(2), 203, 1996.
118. Lagord, C., Berry, M., and Logan, A. Expression of TGF[beta]2 but Not TGF[beta]1 correlates with the deposition of scar tissue in the lesioned spinal cord. *Molecular and Cellular Neuroscience* 20(1), 69, 2002.
119. Asher, R. A., Morgenstern, D. A., Fidler, P. S., Adcock, K. H., Oohira, A., Braistead, J. E., Levine, J. M., Margolis, R. U., Rogers, J. H., and Fawcett, J. W. Neurocan is upregulated in injured brain and in cytokine-treated astrocytes. *Journal of Neuroscience* 20(7), 2427, 2000.
120. Logan, A., Green, J., Hunter, A., Jackson, R., and Berry, M. Inhibition of glial scarring in the injured rat brain by a recombinant human monoclonal antibody to transforming growth factor-beta2. *European Journal of Neuroscience* 11(7), 2367, 1999.
121. Ferguson, M. and Leigh, I. Wound healing. In Champion, R. (ed), *Rook/Wilkinson/Ebling Textbook of Dermatology*, Vol. 1. Oxford, U.K., Blackwell Science, 1998, p. 337.
122. Wolff, J. *Das Gestz der Transformation der Knochen*. Berlin, Hirschwald, 1892.
123. Parfitt, A. Osteonal and hemi-osteonal remodeling: the spatial and temporal framework for signal traffic in adult human bone. *Journal of Cellular Biochemistry* 55, 273, 1994.

124. Burnett, M. and Zager, E. Pathophysiology of peripheral nerve injury: a brief review. *Neurosurgical Focus* 16(5), 1, 2004.
125. Bellamkonda, R. V. Peripheral nerve regeneration: an opinion on channels, scaffolds and anisotropy. *Biomaterials* 27(19), 3515, 2006.
126. Ehrlich, H. and Rajaratnam, J. Cell locomotion forces versus cell contraction forces for collagen lattice contraction: An *in vitro* model of wound contraction. *Tissue and Cell* 22(4), 407, 1990.
127. Tomasek, J. J., Gabbiani, G., Hinz, B., Chaponnier, C., and Brown, R. A. Myofibroblasts and mechano-regulation of connective tissue remodeling. *Nature Reviews Molecular Cell Biology* 3(5), 349, 2002.
128. Gong, M. C., Iizuka, K., Nixon, G., Browne, J. P., Hall, A., Eccleston, J. F., Sugai, M., Kobayashi, S., Somlyo, A. V., and Somlyo, A. P. Role of guanine nucleotide-binding proteins—ras-family or trimeric proteins or both—in Ca2+ sensitization of smooth muscle. *Proceedings of the National Academy of Sciences USA* 93(3), 1340, 1996.
129. Hinz, B., Mastrangelo, D., Iselin, C. E., Chaponnier, C., and Gabbiani, G. Mechanical tension controls granulation tissue contractile activity and myofibroblast differentiation. *American Journal of Pathology* 159(3), 1009, 2001.
130. Desmouliere, A., Redard, M., Darby, I., and Gabbiani, G. Apoptosis mediates the decrease in cellularity during the transition between granulation tissue and scar. *American Journal of Pathology* 1, 56, 1995.
131. Jee, W. S. S., Mori, S., Li, X. J., and Chan, S. Prostaglandin E2 enhances cortical bone mass and activates intracortical bone remodeling in intact and ovariectomized female rats. *Bone* 11(4), 253, 1990.
132. Fan, X., Roy, E., Zhu, L., Murphy, T. C., Ackert-Bicknell, C., Hart, C. M., Rosen, C., Nanes, M. S., and Rubin, J. Nitric oxide regulates receptor activator of nuclear factor-{kappa}B ligand and osteoprotegerin expression in bone marrow stromal cells. *Endocrinology* 145(2), 751, 2004.
133. Spataro, L., Dilgen, J., Retterer, S., Spence, A. J., Isaacson, M., Turner, J. N., and Shain, W. Dexamethasone treatment reduces astroglia responses to inserted neuroprosthetic devices in rat neocortex. *Experimental Neurology* 194(2), 289, 2005.
134. Linkins, L. and Weitz, J. New anticoagulant therapy. *Annual Review of Medicine* 56, 63, 2005.
135. McGraw, J., Hiebert, G., and Steeves, J. Modulating astrogliosis after neurotrauma. *Journal of Neuroscience Research* 63(2), 109, 2001.
136. Sofroniew, M.V. Reactive astrocytes in neural repair and protection. *Neuroscientist* 11(5), 400, 2005.

Part II

2 Signal Considerations for Chronically Implanted Electrodes for Brain Interfacing

Warren M. Grill

CONTENTS

2.1 INTRODUCTION

2.1.1 INFORMATION EXCHANGE WITH THE NERVOUS SYSTEM

The fundamental unit of communication within the nervous system is the action potential. The action potential is an electrochemical signal mediated by the flow of ions across the membrane of individual neurons between the intracellular and extracellular spaces. This ionic flux results from concentration and potential gradients between the intracellular and extracellular spaces and occurs through ion-permeant pores (transmembrane proteins or channels) present in the membrane. Generation of action potentials by electrical stimulation or transduction of action potentials by electrical recording forms the basis of neural prosthetic interfaces.

A nerve cell or a nerve fiber can be artificially stimulated by depolarization of the cell's membrane. The resulting action potential propagates to the terminal of the neuron, leading to release of neurotransmitter that can impact the postsynaptic cell. Passage of current through extracellular electrodes positioned near neurons creates extracellular potentials in the tissue, and the potential distribution can result in depolarization and generation of an action potential. Alternately, extracellular potentials may modulate or block ongoing neuronal firing depending on the magnitude, distribution, and polarity of the potentials. Generation or modulation of action potentials with applied currents is the basis for neural prosthetic devices that use electrical stimulation for restoration of function.

Conversely, the flow of ionic current gives rise to potentials within the tissue, and these potentials can be recorded to extract information from the nervous system. It is possible to record the activity from single neurons, multiunit activity from several active neurons, or so-called field potentials resulting from the activity of large populations of neurons. The recording of the electrical potentials generated by single or aggregate action potentials is the basis for recovering command or control signals directly from the nervous system for use in control of external devices.

2.1.2 EXAMPLES OF BRAIN DEVICES

Nominally, a neural interface for stimulation or recording of the central nervous system includes an implanted electrode or electrode array, lead wire, and in many instances an implanted electronics package (Figure 2.1A). Examples of application of central nervous system stimulation from treatment of neurological disorders include the treatment of pain by stimulation of the brain [1] and spinal cord [2], treatment of tremor and the motor system symptoms of Parkinson's disease [3], as an experimental treatment for epilepsy [4,5], as well as a host of other neurological disorders [6]. In addition, central nervous system (CNS) stimulation is being developed for restoration of hearing by electrical stimulation of the cochlear nucleus [7] and for restoration of vision [8–10]. Examples of the spectrum of electrode types used for CNS stimulation (and recording) are shown in Figure 2.2.

2.1.3 CONSIDERATION OF NEURONAL DAMAGE

It is critical to understand the nature and mechanisms of the tissue response to the implantation, residence, and in the case of stimulation, activation of electrodes in the CNS (Figure 2.1B). These devices are rapidly becoming more widespread,

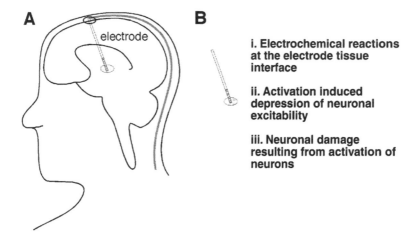

A electrode

B

i. Electrochemical reactions at the electrode tissue interface

ii. Activation induced depression of neuronal excitability

iii. Neuronal damage resulting from activation of neurons

FIGURE 2.1   Elements of a system for brain interfacing. (A) Implanted components include an electrode or electrode array, lead wire or lead cable, and either a percutaneous connection or an implanted electronics package. The implanted electronics may be a pulse generator for instances of brain stimulation or signal conditioners (amplifiers, filters) and signal telemeters for instances of brain recording. (B) The tissue response around the electrode(s) can result from passive mechanical or chemical factors, from activity dependent changes in the excitability of neurons, and/or from the products of electrochemical products at the electrode–tissue interface, and can lead to changes in device functionality.

smaller, and more dense. Unfortunately, there remains a lack of first principles understanding of the mechanisms of neuronal injury. Thus, the issue of damaging versus nondamaging neural interfaces has been and will continue to be addressed in a purely empirical manner.

Analysis of postmortem human and animal tissue has shown that there is neuronal loss around chronically implanted electrodes and a high density of astrocytes, microglia, and vasculature around the electrode. Loss of neurons around the electrode may affect how well the neural prosthesis functions, especially as devices move toward smaller arrays of electrodes that use microstimulation or single-unit recording of small numbers of single neurons.

2.2   BIOPHYSICAL CONSIDERATIONS

The efficacy of neural interfaces for neural stimulation and recording is strongly dependent on the physical proximity of the electrode to the neurons targeted for stimulation or recording. The tissue response to electrode implantation, residence, and activation can increase this distance through the formation of a connective tissue and glial scar sheath around the electrode and through death of neurons in close proximity to the electrode [11,12].

2.2.1   THE IMPORTANCE OF PROXIMITY FOR STIMULATION

The current required for extracellular stimulation of neurons, threshold, I_{th}, increases as the distance between the electrode and the neuron, r, increases. This is described by the current–distance relationship [13]:

$$I_{th} = I_R + k \cdot r^2$$

where the offset, I_R, determines the absolute threshold, and the slope, k, determines the threshold difference between neurons at different distances from the electrode. Current distance relationships for excitation of axons and cells in the CNS have been measured in many preparations, and current distance curves for these two populations are summarized in Figure 2.3. Any increase in the electrode to neuron distance, as a result of the tissue response, will increase the current required for excitation. Subsequent increases in stimulation intensity, as required to achieve the desired effect, may exacerbate the tissue response, leading to a further increase in the electrode to neuron distance, and so on in a positive feedback fashion.

2.2.2 EFFECT OF NEURONAL DEATH ON STIMULATION INPUT–OUTPUT PROPERTIES

The dependence of the neuronal excitation threshold on the electrode to neuron distance, neuron size, and neuron excitability gives rise to a distribution of thresholds across a population of neurons (Figure 2.4A). Therefore, input–output or recruit-

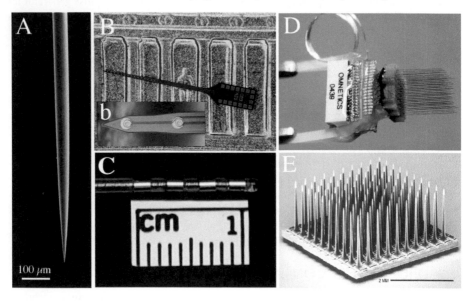

FIGURE 2.2 Examples of electrodes for brain stimulation and recording. (A) Single iridium microwire electrode developed at Huntington Medical Research Institute that can be used for extracellular recording from single units or extracellular microstimulation of small populations of neurons. (Image courtesy of D. B. McCreery, Huntington Medical Research Institute [31].) (B) Multisite silicon microprobe developed at the University of Michigan and higher magnification view (b) of two electrode sites near the tip. (Images courtesy of J. F. Hetke, University of Michigan.) (C) Quadrapolar electrode used for deep brain stimulation (Medtronic Inc., Minneapolis, Minnesota). (D) Arrays of up to 128 microwires enable simultaneous extracellular recording from multiple single neurons. Each wire is 50 μm diameter stainless steel, insulated with Teflon. (Image courtesy of Miguel A. L. Nicolelis, Duke University [55].) (E) Multielectrode silicon array developed at the University of Utah. (Image courtesy of R. A. Normann et al. [56].)

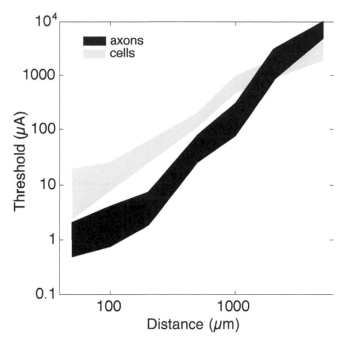

FIGURE 2.3 Effect of electrode to neuron distance on stimulation intensity thresholds. The current–distance relationship describes the threshold intensity required for stimulation as a function of the distance between the electrode and the neuron. Current distance curves for axons and cells were constructed from data summarized in Ranck, 1976 [57].

ment curves of the number of neurons stimulated (or, equivalently, the amplitude of the response mediated by neuronal activation) as a function of the stimulation intensity are sigmoidal (Figure 2.4B). Damage or death of neurons around the electrode can give rise to changes in the input–output function (i.e., impact the response evoked by electrical stimulation).

We conducted computer simulations to quantify the effects of neuronal death on the input–output properties of neural stimulation. Simulations were conducted under two conditions. First, we assessed changes that might result when neurons die as a function of their proximity to the stimulating electrodes. This would occur, for example, when the tissue reaction to the electrode causes neuronal death or if electrochemical by-products, produced at the electrode–tissue interface, were harmful to the surrounding neurons. Second, we assessed changes in input–output properties that occurred when neurons died as a function of their excitability, that is, when the most excitable neurons died first. This situation would occur when neuronal death resulted from the induced activity, for example, as a result of metabolic overload.

In all cases, neuronal death resulted in an increase in the threshold stimulation intensity as well as a reduction in the amplitude of the response (number of neurons activated) for a given stimulation current. These effects were observed whether considering a population of axons (Figure 2.4C) or a population of local cells (Figure 2.4D). Killing neurons according to their relative excitability produced proportionally larger shifts in threshold, and the foot of the input–output curve shifted to

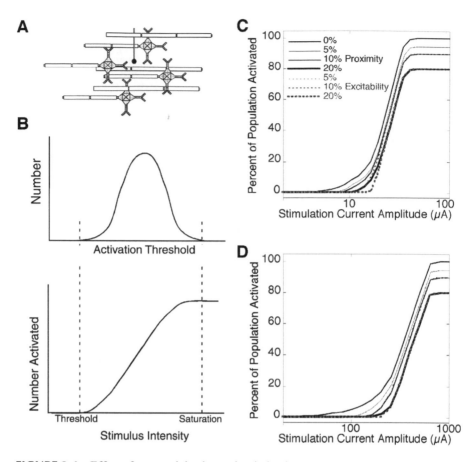

FIGURE 2.4 Effect of neuronal death on stimulation input–output properties. (A) A population of neurons distributed at different distances and positions with respect to a stimulating
electrode. (B) The distribution of neuronal positions and excitability yields a distribution
of thresholds to excite individual neurons. The recruitment (input–output) curve describes
the number of neurons stimulated as a function of the stimulus intensity, where the intensity can be modulated through changes in either the amplitude or duration of the stimulus
pulse. The sigmoidal shape of the recruitment curve arises from cumulative integration of
the distribution of thresholds as the stimulation intensity is increased. (C) Computation of
the effect on the input–output properties of a population of nerve fibers following death of
different proportions of neurons according to either their proximity to the stimulating electrode or their excitability. (D) Computation of the effect on the input–output properties of
a population of local cells following death of different proportions of neurons according to
either their proximity to the stimulating electrode or their excitability. The simulations in (C)
and (D) included 500 neurons distributed uniformly in a 600 μm x 600 μm x 600 μm cube
with a point source electrode at the center. The tissue was an infinite isotropic homogeneous
volume with resistivity 500 Ω-cm. Stimuli were monophasic rectangular 0.1 ms duration
cathodic pulses, and threshold was determined to ± 0.1 % by variable time-step numerical
integration. The neuron models were based upon the S-type motoneuron model in McIntyre
and Grill, 2000 [58].

the right. The difference between the control input–output curve and the curve with neurons killed according to their excitability was approximately constant over the full range of recruitment. In contrast, the input–output curves of neuronal populations with neurons killed according to their proximity to the electrode were closer to the control curve near the foot of the curve and deviated more as the activation level increased to converge with the curve of neurons killed by excitability. Overall, the input–output curves were very similar whether neurons died according to their proximity to the electrode or according to their excitability. These results suggest that neuronal death according to either criterion is likely to have a similar effect on device function and that measurement of changes in input–output curves is not likely to be a sensitive diagnostic for the characteristics of neuronal death.

2.2.3 IMPACT OF ELECTRODE TO NEURON DISTANCE ON RECORDED SIGNALS

As with stimulation intensity, the amplitude of the signal recorded from an active neuron (or neurons) varies inversely with the distance between the source (active neurons) and the recording electrode, r. For a monopolar source the potential varies as $1/r$, while for a dipolar source, more typical of transmembrane currents, the potential varies as $1/r^2$.

The historical rule of thumb is that to record isolated signals from single active neurons requires that the electrode lie within $r = 100$ μm of the active neuron. The relationship between the recorded signal amplitude and the estimated electrode to neuron distance obtained from two published studies were replotted. These data (Figure 2.5) are consistent with this rule of thumb and indicate that the recorded signal amplitude falls off rapidly as the electrode to neuron distance is increased.

2.3 MECHANISMS OF NEURONAL INJURY AND DEATH

Given the clear performance consequences of increases in the electrode to neuron distance on both stimulation thresholds and recorded signal amplitudes, considerations of the underlying contributors to neuronal injury or death (Figure 2.1B) are of paramount importance in designing neural prosthetic interfaces. Most generally, the tissue response can be classified as either passive (resulting from the presence of the electrode) or active (resulting from the passage of stimulus current), although it is not always possible to differentiate between these two contributions to the tissue response. The passive tissue response can result from surgical trauma and the mechanical and chemical properties of the implant. The active tissue response results from electrochemical reaction products formed at the tissue–electrode interface and from physiological changes associated with neural activity.

2.3.1 PASSIVE PROPERTIES

The tissue response to an implanted electrode depends on the physical attributes of the device [14] including its size, shape [15–17], and surface texture [18]. In addition, the chemical or material composition of the device has a strong impact on the tissue response to the material and ranges from the benign (so-called biocompatible materials) to those that are toxic and lead to cell death [19–21].

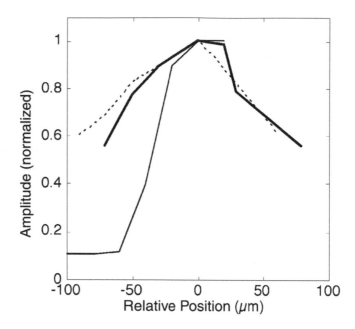

FIGURE 2.5 Effect of electrode to neuron distance on recorded potential amplitudes. Amplitude of extracellular action potentials recorded from single neurons as a function of the distance of the recording electrode from the point of maximum amplitude. Each line represents a different single unit recording. (Data extracted from Figure 7 in Drake et al., 1988 [59] and Figure 8 in Cham et al., 2005 [60].)

The "passive" tissue response to the residence of the electrode within the brain can result from ongoing mechanical injury to the tissue as a result of electrode motion [22], as well as from the electrode materials. Even in quiescent, anesthetized animals, intrinsic brain movement from respiration exceeds 10 μm [23], and brain displacement may be much larger during changes in posture and free movement. Such motion can lead to variability in the recorded signal (Figure 2.5) or stimulation-evoked response (Figure 2.3) and lead to exacerbation of the tissue response to the residence of the device. There is a substantial mismatch between the mechanical properties of electrode materials and brain tissue, and this leads to substantial strain at the electrode–brain interface [22]. The shear modulus of the brain is 200 to 1500 Pa [24,25], which is several orders of magnitude less than the shear modulus of silicon (~50 GPa) or tungsten (~130 GPa), two commonly used electrode materials. Recall that shear modulus, G, is related to Young's modulus, E, and the Poisson ratio, v, by $G = E/[2*(1 + v)]$. Several efforts have been made to develop polymer-based electrodes, but there is still a substantial mechanical mismatch between the properties of the brain and the properties of commonly used polymeric substrates including polyimide (~1 GPa), liquid crystal polymer (~3 GPa), and parylene (~1 GPa). Computational studies suggest that tighter mechanical coupling between the electrode and the brain [22], for example, through the use of integrative coatings, can reduce the strain at the electrode–tissue interface [26].

2.3.2 ACTIVE PROPERTIES

The tissue response due to the passage of current as required for neuronal stimulation may arise from electrochemical reaction products formed at the electrode-tissue interface or from physiological changes in the tissue that are associated with neural excitation. It is difficult to generalize the results obtained in studies of tissue damage resulting from electrical stimulation to new electrode geometries or materials, new stimulation patterns, or different anatomical regions. As described by Agnew and colleagues "the manner and degree to which this [neural injury] occurs depends upon the physical and pharmacological composition of the particular neural substrate, as well as the stimulus parameters" [27]. The tissue response to the active properties of a neural interface can result from electrochemical reaction products formed at the tissue–electrode interface and from physiological changes associated with neural activity, which include both transient depression of excitability and neuronal damage. Each of these three active effects is considered in turn.

2.3.2.1 Electrochemistry of the Electrode–Electrolyte Interface

Electrical stimulation is typically delivered using metal electrodes, which carry current as the flow of electrons, implanted in the body, which carries current as the flow of ions (Figure 2.6A). Thus, there exists an interface between the metal electrode and the ionic conductor of the body. In general this interface has a nonlinear impedance that is a function of the voltage across the interface. This interface impedance can impact the properties of stimulation, and electrochemical reactions at the electrode–tissue interface can lead to electrode dissolution and production of chemical species that may be damaging to tissue.

Reactions that take place at the electrode–tissue interface are dependent on the potential of the electrode as well as the time course of the stimulus pulse and may be accelerated by increased current density. The electrode interface can be modeled by the parallel combination of a capacitor (C), representing the double-layer capacitance, and nonlinear impedances representing faradaic electrochemical reactions (Figure 2.6B). The voltage developed across the electrode–tissue interface (V) determines which chemical reactions will take place and is dependent on the amount of charge in the stimulus pulse (Q = current intensity * pulse duration) and the electrode capacitance, since $V = Q/C$. The electrode capacitance is determined by the properties of the material and is proportional to the electrode area, A. Therefore, the potential developed across the interface is proportional to electrode area. This relationship is the basis for the correlation between charge density and tissue damage and the assertion that the charge density is an indirect measure of the electrochemical contribution to tissue damage (see below).

If the voltage across the electrode–tissue interface is kept within certain limits, then chemical reactions can be avoided, and all charge transfer will occur by the charging and discharging of the double-layer capacitance [21,28]. However, in many instances, the electrode capacitance is not sufficient to store the charge necessary for the desired excitation without the electrode voltage reaching levels where reactions will occur. The principal approach to control the interface voltage has been the use of charge-balanced biphasic stimuli that have two phases that contain equal and opposite charge [29,30]. However, even with charge-balanced pulses it is pos-

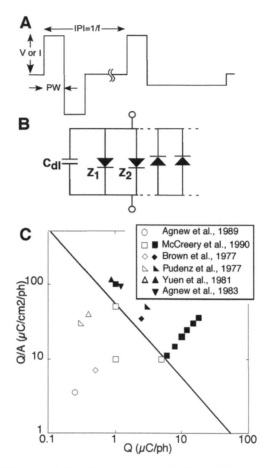

FIGURE 2.6 (A) Characteristics of waveforms typically used for electrical stimulation of the nervous system. Typical stimuli are biphasic-balanced charge pulses with a duration of 0.02 to 0.5 ms/phase, an amplitude of 0.1 to 10 mA, and a frequency of 10 to 200 Hz. (B) Equivalent circuit models of a metal electrode in an ionic conducting medium. C_{dl} is the double layer capacitance and the Z_i represent potential-dependent electrochemical reactions that can occur at the electrode–tissue interface. (C) Relationship between charge per phase and charge density for damaging and nondamaging electrical stimulation. Charge is the product of current and pulse width, and charge density is obtained by dividing by the electrode area. The damaging and nondamaging regions are based upon previous experimental studies. (Data compiled from Figure 9.1 of Agnew et al., 1990 [11] and Figure 8 from Merrill et al., 2005 [21].) The line is the Shannon model with k = 1.75.

sible that the interface voltage may reach levels where electrochemical reactions can occur.

2.3.2.2 Activity-Dependent Changes in Neuronal Excitability

During artificial electrical stimulation, relatively large numbers of neurons are active synchronously. This nonphysiological pattern of activation results in activity-dependent changes in neuronal excitability [31], as well as apparent activity-dependent neuronal injury.

Continuous stimulation of large-diameter peripheral nerve fibers can cause a prolonged increase in their electrical threshold. Eight hours of 50 Hz stimulation of the common peroneal nerve at suprathreshold intensities produced a reduction in the amplitude of the evoked compound nerve action potential in the large-diameter nerve fibers [32,33]. In some instances this reduction in excitability recovered after 7 days of no stimulation, even in nerves that exhibited axonal degeneration, while in others it did not. The reduction in excitability was mitigated by using intermittent duty cycles and lower frequencies (20 Hz) of stimulation. Similarly, stimulation of the cochlear nerve can produce reversible elevations in electrical thresholds [34].

Continuous stimulation of the cortex also produced reversible increases in the thresholds to produce both direct activation as well as trans-synaptic activation of cortical pyramidal tract neurons [35]. Stimulation at 20 Hz for 24 h or 23 h/day for 1 week with suprathreshold pulses (40 to 80 μA, 200 μs per phase, 400 to 800 μC/cm^2) generated profound but reversible decreases in neuronal excitability. These elevations in threshold were not associated with damage or loss of neurons associated with the delivery of stimulation [36]. However, with the stimulation intensity increased to 320 μA (3200 μC/cm^2), depression of excitability following 24 h of continuous stimulation was not reversed by 7 to 12 days after stimulation. In two instances this was associated with highly localized neuronal damage around the tips of platinum–iridium electrodes but not around other platinum–iridium or activated iridium microelectrodes. Thus, there was not a clear correlation between either reversible or irreversible increases in electrical threshold and neuronal damage. Furthermore, stimulus levels that when applied to a single microelectrode did not produce depression of excitability, when applied simultaneously to multiple electrodes within an array did induce substantial depression of excitability. This supports for the concept that changes in excitability are driven through some "mass action" arising from the simultaneous activation of a critical number of neurons. Electrical microstimulation of the cochlear nucleus, as might be used for restoration of hearing, also produces transient reductions in neuronal excitability [37].

2.3.2.3 Stimulation-Induced Neuronal Injury

The number and spatial extent of neurons activated are related to the charge injected in each stimulus pulse, while the charge density (charge per unit area of the stimulating contact) contributes to the type and rate of electrochemical reactions that occur at the electrode tissue interface. The finding that charge per phase and charge density are both factors in determining the threshold for neural injury by stimulation in both the peripheral nervous system (PNS) [32] and the CNS [38] suggests that in addition to electrochemical mechanisms, physiological mechanisms also contribute to neural damage.

The correlation of charge per phase and tissue damage is thought to result from physiological changes associated with neural excitation. Physiological changes resulting from synchronous activation of a population of neurons increase as the number of excited neurons increases. This relationship may create the correlation between charge per phase and neural damage. There is substantial support for this hypothesis in the PNS and the CNS. In the PNS, peripheral nerve injury can still occur at very low charge densities [32], and anesthetic block of electrical activity occludes the damaging effect of stimulation [39]. In the CNS, equivalent tissue damage was seen under platinum electrodes and tantalum pentoxide capacitor-type electrodes (which were presumed not to have electrochemical reactions) [40]. These studies, which have attempted to differentiate electrochemically induced and activity-induced tissue damage, support the finding that tissue injury results from physiological changes in the neural environment resulting from synchronous activation of populations of neurons.

2.3.2.4 The Shannon Model

The experimental data on the role of charge and charge density in determining whether a stimulation condition is likely to cause tissue damage are well described by the quantitative model of Shannon [41]. This model was originally conceived to describe the empirical data of McCreery et al. [38] from cortical stimulation with circular disk electrodes, but the model is also consistent with other seemingly disparate data. The Shannon model is described using the equation, $\log(Q/A) = k - \log(Q)$, where Q is the charge per phase of the stimulus pulse, A is the electrode area, and k is a constant derived from empirical data to define the boundary between stimulus parameters that produced tissue damage and those that did not (Figure 2.6C).

2.4   A CASE STUDY: DEEP BRAIN STIMULATION

Deep brain stimulation (DBS) has emerged rapidly as a successful treatment for the symptoms of movement disorders including Parkinson's disease, essential tremor, and dystonia and is under investigation for treatment of epilepsy as well as psychiatric disorders including depression and obsessive–compulsive disorder. The DBS system includes an electrode that is surgically implanted within a brain target, according to the disease being treated, connected by a lead extension to a battery-powered pulse generator implanted in the chest. The electrode leads used for DBS (Model 3387 and 3389, Medtronic, Inc., Minneapolis, Minnesota) are cylindrical with macroscopically smooth texture. The four electrode conductors are platinum alloyed with iridium, and the insulating jacket is polyurethane (Figure 1.2C). The effects of treatment are typically only realized when the device is on, and thus this is a permanently implanted medical device intended to last the lifetime of the recipient.

2.4.1   PASSIVE PROPERTIES

There is evidence for a strong passive effect due to DBS electrode insertion in the brain. Immediately after implantation, in many cases the symptoms disappear because of a transient lesioning effect of device placement, most often attributed to edema around the electrode. There are also maintained insertional effects, apparently caused by the lesion made by the presence of the electrode. This has been observed in pain treatment [42] as well as in epilepsy treatment [5]. However, for the treatment of movement disorders, the continued delivery of stimulation is required for the maintenance of effect, and the effect of the implantation-associated lesion is resolved within weeks following implantation.

2.4.2 ACTIVE PROPERTIES

The typical output parameters for DBS are 3 V stimulation amplitude (i.e., 3 mA into a nominal 1 kΩ load), 90 µs pulse duration, and 100 to 190 Hz stimulation frequency (Figure 1.6A), and the devices are typically on for 12 to 24 hours per day.

2.4.2.1 Preclinical Studies

Preclinical studies were conducted by Medtronic, Inc. to evaluate the tissue response to DBS lead implantation and stimulation [43]. Acute studies were conducted in 10 pigs implanted bilaterally with Model 3387 leads and stimulated for 7 hours at supraclinical parameters (10.5 V, 130 to 185 Hz, and pulse durations from 0.913 to 2.0 ms). Postmortem pathology revealed tissue damage around both unstimulated and stimulated leads, and the difference was detectable in only 3 of 9 cases. The tissue response included inflammation, gliosis, and neuronal degeneration. Subsequently, a chronic study was conducted in 8 pigs implanted bilaterally with Model 3387 leads for 2 to 9 months of continuous stimulation with parameters more similar to those used clinically (185 Hz, pulse duration of 0.45 ms, and voltage set to maximum tolerable). From the 14 lead pairs that were evaluated, 11 (79%) showed differential effects of stimulation and 3 (21%) did not. The tissue response within 2 mm of the lead included inflammation, mineralization, gliosis, and neuronal degeneration.

2.4.2.2 Autopsy Findings

A number of studies have examined the tissue response to chronically implanted DBS electrodes. This review is restricted to DBS for the treatment of movement disorders using electrodes (Models 3387, 3389) that have well-characterized materials and regulatory approval.

Caparros-Lefebvre reported on a patient who was discontinuously stimulated over the course of 43 months [44] with DBS of the Vim thalamus to treat Parkinsonian tremor. Gliosis and vacuolization were observed within 1 mm of the electrode, and adjacent to polyurethane insulation macrophages and giant cells were observed. The authors concluded that the "lesion" resulting from implantation and application of stimulation was smaller than functional thalamic lesions placed to treat motor symptoms.

Haberler and colleagues [45] reported on a series of 8 patients with Parkinson's disease who had received either thalamic (6 cases) or subthalamic nucleus (STN) (2 cases) DBS for up to 70 months. The brain areas around 11 leads were examined following between 3 and 70 months of stimulation, and the response to a single lead was examined 2 days after implantation. The leads were surrounded by a fibrous tissue capsule 5 to 25 µm thick, followed by a rim of a glial fibrillary acidic protein

(GFAP)-positive gliosis about 500 μm thick, followed by a 1-mm thick annulus of scattered GFAP-reactive astrocytes. Furthermore, mononuclear leukocytes, macrophages, and multinucleated giant cells were observed in the track and surrounding tissue. In one patient the giant cells were seen to have engulfed fragments of foreign material. No loss of axons or demyelination was observed, and healthy neurons were found immediately adjacent to the electrode tracks.

Two observations from this study suggest that stimulation per se did not contribute to the observed tissue response. First, the tissue characteristics were similar across cases, independent of the duration of stimulation. Second, the tissue characteristics were the same in the vicinity of insulated and on stimulated regions of the lead as over the stimulated portions of the lead. The authors concluded that "chronic DBS does not cause damage to adjacent brain parenchyma" [45].

In a case report, Henderson et al. [46] reported the presence of a thalamic lesion following implantation and stimulation for 2 years to treat Parkinson's disease. The electrode tract was surrounded by a glial sheath and tissue vacuoles, and the tip was located within the centromedian–parafascicular nucleus of thalamus (CM/Pf). The tip was surrounded by an apparent lesion and increased astrocytosis, and there was extensive neuronal loss within the CM/Pf. The extensive lesion present in this case differs from the benign tissue response described for most other cases, although the stimulus parameters were quite similar, and the authors conclude that DBS may be associated with thalamic damage. It is noteworthy that this electrode was implanted to replace a lead initially placed within the Vim thalamus and forcibly removed by the patient.

In a second case report, Henderson et al. [47] described the tissue response to STN DBS in a patient with Parkinson's disease who died 2 months after implantation. An inflammatory reaction was present within 1 mm of the electrode tract, characterized by astrogliosis, the appearance of macrophages and microglia, and GFAP staining. Furthermore, there was apparent "minor cell loss" within the thalamus through which the tract passed as well as in the STN. The authors concluded that little tissue damage is associated with STN DBS.

In another case report, Burbaud et al. [48] reported on the tissue response to 2 years of DBS in the Vop thalamus in a patient with chorea–acanthocytosis. The right electrode was forcibly removed by the patient and was subsequently replaced at the same coordinates. The corresponding electrode tract exhibited astrocytic gliosis, infiltration of lymphocytes, and iron deposits. The presence of iron around the implant site is somewhat perplexing, but this could be the result of hemosiderin containing macrophages [45]. The left electrode tract induced less tissue response, and at the tip of both electrodes there was focal gliosis and leukocyte infiltration. The authors concluded that DBS induced only minor changes in the tissue in the vicinity of the electrode.

Henderson and colleagues reported on a third case of a patient with Parkinson's disease who received DBS of the Vim thalamus for 7 years [49]. A region of astrocytosis surrounded the electrode track. Within 0.5 mm of the stimulated electrode and the electrode tip there was astrocytosis, an apparent reduction in the density of myelin, and significant loss of neurons (60% loss compared to the contralateral Vim), and the remaining neurons appeared pyknotic. However, there were no significant changes in neuronal density near the electrodes not used for stimulation or in the region further (0.5 to 1.5 mm) from the active electrode. The authors also elaborated on their earlier report [47] to reveal that neuronal loss within the STN was approximately 30% and was noted within 0.5 mm of the stimulated contact but not near the unstimulated contacts. These results suggest, and the authors concluded, that the neuronal loss may have been due to stimulation, rather than just the surgical introduction of the electrode. Importantly, even with these levels of neuronal loss, the resulting "lesion" was insufficient to treat the movement disorder, and efficacy required the continued delivery of stimulation.

In an abstract reporting a single case of DBS in the STN to treat Parkinson's disease, Larsen et al. [50] described that the electrode track was surrounded by a GFAP-positive "capsule" approximately 150 µm thick constituted of a "thin collagen layer lining the lumen of the tract" surrounded by a region of decreased neuronal density. There were no apparent differences in tissue in regions receiving stimulation and regions of no stimulation.

Moss et al. [51] reported on the characteristics of the tissue that was adhered to DBS electrodes removed after 3 to 31 months for clinical reasons including suboptimal positioning (14/21 leads), scalp infection (2/21), increased impedance (2/21), fracture of the electrode (1/21), damaged connection (1/21), or displacement of the electrode (1/21). Scanning and transmission electron microscopy revealed the presence of multinucleated giant cells and mononuclear macrophages. Electron-dense material was present within and around both cell types, suggesting active phagocytosis, and although it was suggested that this material might be polyurethane particles, subsequent microscopy of polyurethane insulation did not support this conclusion.

2.4.2.3 Charge Density Limits

The Shannon model, and the data on which it was based, provided the rationale for the manufacturer of DBS systems to recommend that stimulus parameters be selected not to exceed a charge density of 30 µC/cm². Since the electrode area is fixed (0.059 cm² for the 3387 and 3389 leads), this leads to a line relating charge per phase to charge density (Figure 2.7). However, several factors must be considered when evaluating this recommendation, and the human postmortem studies that reported substantial loss of neurons in proximity to the active electrode [49] employed parameters well below this limit.

First, all human postmortem studies to date are from cases where stimulus parameters were set well under the proposed limit, so the lack of pathology in these cases does not necessarily support this limit. Second, the stimulation frequency used to generate the data from which the limit was established, was in some cases lower than the ≥130 Hz frequency typically employed with DBS, and the appearance of neural damage appears to have a frequency-dependent threshold. Finally, the distribution of charge across the electrode contact is highly nonuniform [52],

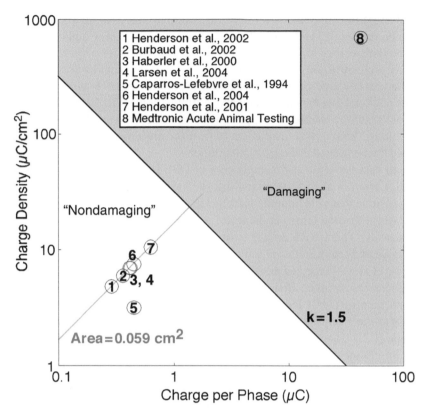

FIGURE 2.7.   Relationship between charge per phase and charge density for damaging and nondamaging electrical stimulation, as relates to deep brain stimulation. The damaging and nondamaging regions are based upon previous experimental studies and described using the Shannon model with k = 1.5. The combinations of charge and charge density from postmortem studies of deep brain stimulation (DBS) are indicated by the circles. The clinical DBS electrode has an area of 0.059 cm² and results in the gray line.

and the peak charge density is likely to exceed the recommended limit, even when the average charge density does not.

2.4.2.4   Performance Considerations

These studies indicate that the implantation and residence of DBS electrodes evokes a classic chronic foreign body response characterized by the presence of a fibrous tissue capsule, often termed a "glial scar," surrounded by a region of gliosis, as demonstrated by upregulation of GFAP. The spatial extent of the resulting scar, and in some cases loss of neurons, extended further than 1 mm from the electrode. This apparent increase in the electrode to neuron distance is expected to increase the stimulation intensity required for effect (Figure 2.3) relative to that which would be required with a less pronounced tissue response. However, in all cases the resulting lesion was insufficient to treat the movement disorder, and efficacy required continued stimulation, and there are no reported decreases in device performance as a result of tissue damage.

The lack of correlation between the level of stimulation and the tissue response [45] and the lack of difference in tissue adjacent to stimulated and unstimulated electrodes [50] appear at odds with the reports of Henderson and colleagues [46, 47, 49] indicating significant neuronal loss near the stimulated electrodes but not near unstimulated electrodes. Two factors may contribute to these disparate results. First, the quantitative neuronal analysis used by Henderson et al. was likely more sensitive than the qualitative observations of others, and the latter may have missed reductions in neuronal density. Second, some neuronal loss may have been mechanical in nature and occurred preferentially at the end of the electrode [53].

2.5 FUTURE CONSIDERATIONS

The chronic application of electrical stimulation of the brain or spinal cord, as required for neural prosthetic devices, can cause damage to the electrode or the underlying tissue. Such damage may render the device ineffective, and thus prevention of such damage is of paramount concern. Clinical implementation of DBS, for example, has not revealed apparent decreases in device performance as a result of tissue damage. However, continued use of these devices in a nondamaging manner requires an understanding of the mechanisms responsible for tissue damage, especially in light of reports of neuronal loss in proximity to active electrodes. As electrodes are made smaller, for example, to improve the selectivity of stimulation, or as stimulus parameters are increased, for example, in the treatment of obsessive compulsive disorder, larger voltages are used more than in the treatment of movement disorders, with the potential risks of tissue damage increase.

The lack of knowledge about the fundamental causes of neuronal injury makes it difficult to determine *a priori* whether a proposed electrode design or stimulation paradigm will be damaging or not. For example, changes in the shape of an electrode will result in changes in the current density distribution on the surface of electrodes. The change in current (or, equivalently, charge) density distribution may affect the propensity for stimulation to generate either tissue damage or electrode damage. Charge density is a factor in determining whether stimuli are damaging or not [38], and electrode corrosion appears to occur preferentially at the edges where the current density is highest [52]. When commenting on studies on neural damage with disk electrodes on the cortical surface [53], Larson [54] pointed out that "...current density and charge density is not uniform beneath the electrode...[and] consequently it is not possible to extend these observations to other electrodes." Thus, until the fundamental causes of neuronal injury are determined, design changes necessitate empirical studies to determine the propensity of a specific geometry or set of parameters to produce neural damage.

ACKNOWLEDGMENTS

Preparation of this chapter was supported by grants R01 NS040894 and R21 NS054048 from the National Institutes of Health. Thank you to Dr. Dustin J. Tyler,

Case Western Reserve University, for information on the mechanical properties of brain tissue and electrode materials, and to Amin Mahnam for conducting the simulations reported in Figure 2.4.

REFERENCES

1. Coffey, R. J. Deep brain stimulation for chronic pain: results of two multicenter trials and a structured review. *Pain Med* 2, 183, 2001.
2. Cameron, T. Safety and efficacy of spinal cord stimulation for the treatment of chronic pain: a 20-year literature review. *J. Neurosurg.* 100(3 Suppl Spine), 254, 2004.
3. Gross, R. E. and Lozano, A. M. Advances in neurostimulation for movement disorders. *Neurol. Res.* 22, 247, 2000.
4. Velasco, M., Velasco, F., and Velasco, A. L. Centromedian-thalamic and hippocampal electrical stimulation for the control of intractable epileptic seizures. *J. Clin. Neurophysiol.* 18, 49, 2001.
5. Hodaie, M., Wennberg, R. A., Dostrovsky, J. O., and Lozano, A. M. Chronic anterior thalamus stimulation for intractable epilepsy. *Epilepsia* 43, 603, 2002.
6. Gross, R. E. Deep brain stimulation in the treatment of neurological and psychiatric disease. *Expert Rev. Neurother.* 4, 465, 2004.
7. Otto, S. R., Brackmann, D. E., Hitselberger, W. E., Shannon, R. V., and Kuchta, J. Multichannel auditory brain stem implant: update on performance in 61 patients. *J. Neurosurg.* 96, 1063, 2002.
8. Brindley, G. S. and Lewin, W. S. The sensations produced by electrical stimulation of the visual cortex. *J. Physiol.* 196, 479, 1968.
9. Schmidt, E. M., Bak, M. J., Hambrecht, F. T., Kufta, C. V., O'Rourke, D. K., and Vallabhanath, P. Feasibility of a visual prosthesis for the blind based on intracortical microstimulation of the visual cortex. *Brain* 119, 507, 1996.
10. Troyk, P., Bak, M., Berg, J., Bradley, D., Cogan, S., Erickson, R., Kufta, C., McCreery, D., Schmidt, E., and Towle, V. A model for intracortical visual prosthesis research. *Artif. Organs* 27, 1005, 2003.
11. Agnew, W. F., McCreery, D. B., Yuen, T. G. H., and Bullara, L. A. Effects of prolonged electrical stimulation of the central nervous system. In *Neural Prostheses: Fundamental Studies,* Agnew, W. F., McCreery, D. B., Eds. Prentice Hall, Englewood Cliffs, NJ, 1990, 225–252.
12. Biran, R., Martin, D. C., and Tresco, P. A. Neuronal cell loss accompanies the brain tissue response to chronically implanted silicon microelectrode arrays. *Exp. Neurol.* 195, 115, 2005.
13. Stoney, S. D. Jr., Thompson, W. D., and Asanuma, H. Excitation of pyramidal tract cells by intracortical microstimulation: effective extent of stimulating current. *J. Neurophys.* 31, 659, 1968.
14. Yuen, T. G. H., Agnew, W. F., Bullara, L. A., and McCreery, D. B. Biocompatibility of electrodes and materials in the central nervous system. In *Neural Prostheses Fundamental Studies*, Agnew, W. F., McCreery, D. B., Eds. Prentice Hall, Englewood Cliffs, NJ, 1990, 198–223.
15. Wood, N. K., Kaminski, E. J., and Oglesby, R. J. The significance of implant shape in experimental testing of biological samples: disc vs. rod. *J. Biomed. Mater. Res.* 4, 1, 1970.
16. Matlaga, B. F., Yasenchak, L. P., and Salthouse, T. N. Tissue response to implanted polymers: the significance of sample shape. *J. Biomed. Mater. Res.* 10, 391, 1976.

17. Edell, D. J., Toi, V. V., McNeil, V. M., and Clark, L. D. Factors influencing the biocompatibility of insertable silicon microshafts in cerebral cortex. *IEEE Trans. Biomed. Eng.* 39, 635, 1992.

18. Taylor, S. R. and Gibbons, D. F. Effect of surface texture on the soft tissue response to polymer implants. *J. Biomed. Mater. Res.* 17, 205, 1983.

19. Loeb, G. E., Walker, A. E., Uematsu, S., and Konigsmark, B. W. Histological reactions to various conductive and dielectric films chronically implanted in the subdural space. *J. Biomed. Mater. Res.* 11, 195, 1977.

20. Stensaas, S. S. and Stensaas, L. J. Histopathological evaluation of materials implanted in the cerebral cortex. *Acta Neuropathol.* 41, 145, 1978.

21. Merrill, D. R., Bikson, M., and Jefferys, J. G. Electrical stimulation of excitable tissue: design of efficacious and safe protocols. *J. Neurosci. Methods* 141, 171, 2005.

22. Lee, H., Bellamkonda, R. V., Sun, W., and Levenston, M. E. Biomechanical analysis of silicon microelectrode-induced strain in the brain. *J. Neural Eng.* 2, 81, 2005.

23. Gilletti, A. and Muthuswamy, J. Brain micromotion around implants in the rodent somatosensory cortex. *J. Neural Eng.* 3, 189, 2006.

24. Miller, K., Chinzei, K., Orssengo, G., and Bednarz, P. Mechanical properties of brain tissue in-vivo: experiment and computer simulation. *J. Biomech.* 33, 1369, 2000.

25. Prange, M. T. and Margulies, S. S. Regional, directional, and age-dependent properties of the brain undergoing large deformation. *J. Biomech. Eng.* 124, 244, 2002.

26. He, W. and Bellamkonda, R. V. Nanoscale neuro-integrative coatings for neural implants. *Biomaterials* 26, 2983, 2005.

27. Agnew, W. F., McCreery, D. B., Yuen, T. G. H., and Bullara, L. A. MK-801 protects against neuronal injury induced by electrical stimulation. *Neuroscience* 52, 42, 1993.

28. Brummer, S. B. and Turner, M. J. Electrochemical considerations for safe electrical stimulation of the nervous system with platinum electrodes. *IEEE Trans. Biomed. Eng.* 24, 59, 1977.

29. Lilly, J. C., Hughes, J. R., Alvord, E. C. Jr., and Galkin, T. W. Brief noninjurious electric waveform for stimulation of the brain. *Science* 121, 468, 1955.

30. Robblee, L. S. and Rose, T. L. Electrochemical guidelines for selection of protocols and electrode materials for neural stimulation. In *Neural Prostheses: Fundamental Studies*, Agnew, W. F. and McCreery, D. B., Eds. Prentice-Hall, Englewood Cliffs, NJ, 1990, 25–66.

31. McCreery, D. B., Yuen, T. G., Agnew, W. F., and Bullara, L. A. A characterization of the effects on neuronal excitability due to prolonged microstimulation with chronically implanted microelectrodes. *IEEE Trans. Biomed. Eng.* 44, 93, 1997.

32. Agnew, W. F., McCreery, D. B., Yuen, T. G., and Bullara, L. A. Histologic and physiologic evaluation of electrically stimulated peripheral nerve: considerations for the selection of parameters. *Ann. Biomed. Eng.* 17, 39, 1989.

33. Agnew, W. F., McCreery, D. B., Bullara, L. A., and Yuen, T. G. H. Effects of prolonged electrical stimulation of peripheral nerve. In *Neural Prostheses: Fundamental Studies,* Agnew, W. F. and McCreery, D. B., Eds. Prentice Hall, Englewood Cliffs, NJ, 1990, 147–167.

34. Tykocinski, M., Shepherd, R. K., and Clark, G. M. Reduction in excitability of the auditory nerve following electrical stimulation at high stimulus rates. *Heart Res.* 88, 124, 1995.

35. McCreery, D. B., Bullara, L. A., and Agnew, W. F. Neuronal activity evoked by chronically implanted intracortical microelectrodes. *Exp. Neurol.* 92, 147, 1986.

36. Agnew, W. F., Yuen, T. G., McCreery, D. B., and Bullara, L. A. Histopathologic evaluation of prolonged intracortical electrical stimulation. *Exp. Neurol.* 92, 162, 1986.

37. McCreery, D. B., Yuen, T. G., and Bullara, L. A. Chronic microstimulation in the feline ventral cochlear nucleus: physiologic and histologic effects. *Heart Res.* 149, 223, 2000.

38. McCreery, D. B., Agnew, W. F., Yuen, T. G., and Bullara, L. Charge density and charge per phase as cofactors in neural injury induced by electrical stimulation. *IEEE Trans. Biomed. Eng.* 37, 996, 1990.

39. Agnew, W. F., McCreery, D. B., Yuen, T. G., and Bullara, L. A. Local anaesthetic block protects against electrically-induced damage in peripheral nerve. *J. Biomed. Eng.* 12, 301, 1990.

40. McCreery, D. B., Agnew, W. F., Yuen, T. G., and Bullara, L. A. Comparison of neural damage induced by electrical stimulation with faradaic and capacitor electrodes. *Ann. Biomed. Eng.* 16, 463, 1988.

41. Shannon, R. V. A model of safe levels for electrical stimulation. *IEEE Trans. Biomed. Eng.* 39, 424, 1992.

42. Hamani, C., Schwalb, J. M., Rezai, A. R., Dostrovsky, J. O., Davis, K. D., and Lozano, A. M. Deep brain stimulation for chronic neuropathic pain: long-term outcome and the incidence of insertional effect. *Pain* 125, 188, 2006.

43. Medtronic, Inc. Summary of Safety and Effectiveness Data Medtronic Activa Tremor Control System. Activa PMA P960009, 1997.

44. Caparros-Lefebvre, D., Ruchoux, M. M., Blond, S., Petit, H., and Percheron, G. Long-term thalamic stimulation in Parkinson's disease: postmortem anatomoclinical study. *Neurology* 44, 1856, 1994.

45. Haberler, C., Alesch, F., Mazal, P. R., Pilz, P., Jellinger, K., Pinter, M. M., Hainfellner, J. A., and Budka, H. No tissue damage by chronic deep brain stimulation in Parkinson's disease. *Ann. Neurol.* 48, 372, 2000.

46. Henderson, J. M., O'Sullivan, D. J., Pell, M., Fung, V. S., Hely, M. A., Morris, J. G., and Halliday, G. M. Lesion of thalamic centromedian–parafascicular complex after chronic deep brain stimulation. *Neurology* 56, 1576, 2001.

47. Henderson, J. M., Pell, M., O'Sullivan, D. J., McCusker, E. A., Fung, V. S., Hedges, P., and Halliday, G. M. Postmortem analysis of bilateral subthalamic electrode implants in Parkinson's disease. *Mov. Disord.* 17, 133, 2002.

48. Burbaud, P., Vital, A., Rougier, A., Bouillot, S., Guehl, D., Cuny, E., Ferrer, X., Lagueny, A., and Bioulac, B. Minimal tissue damage after stimulation of the motor thalamus in a case of chorea-acanthocytosis. *Neurology* 59, 1982, 2002.

49. Henderson, J. M., Rodriguez, M., O'Sullivan, D., Pell, M., Fung, V., Benabid, A. L., and Halliday, G. Partial lesion of thalamic ventral intermediate nucleus after chronic high-frequency stimulation. *Mov. Disord.* 19, 709, 2004.

50. Larsen, M., Bjarkam, C. R., Sørensen, J. C., Bojsen-Møller, M., Sunde, N. A., and Ostergaard, K. Chronic subthalamic deep brain stimulation (STN DBS) in Parkinson's disease: a histopathological study. *FENS Abstr.* 2, A053.3, 2004.

51. Moss, J., Ryder, T., Aziz, T. Z., Graeber, M. B., and Bain, P. G. Electron microscopy of tissue adherent to explanted electrodes in dystonia and Parkinson's disease. *Brain* 127, 2755, 2004.

52. Rubinstein, J. T., Spelman, F. A., Soma, M., and Suesserman, M. F. Current density profiles of surface mounted and recessed electrodes for neural prostheses. *IEEE Trans. Biomed. Eng.* 34, 864, 1987.

53. Yuen, T. G., Agnew, W. F., Bullara, L. A., Jacques, S., and McCreery, D. B. Histological evaluation of neural damage from electrical stimulation: considerations for the selection of parameters for clinical application. *Neurosurgery* 9, 292, 1981.

54. Larson, S. J. Comment on Yuen et al. (1981). *Neurosurgery* 9, 299, 1981.

55. Nicolelis, M. A. L., Dimitrov, D., Carmena, J. M., Crist, R., Lehew, G., Kralik, J. D., and Wise, S. P. Chronic, multisite, multielecrode recording in macaque monkeys. *Proc. Natl. Acad. Sci.* 100, 11041, 2003.

56. Normann, R. A., Maynard, E. M., Rousche, P. J., and Warren, D. J. A neural interface for a cortical vision prosthesis. *Vision Res.* 39, 2577, 1999.

57. Ranck, J. B. Jr. Which elements are excited in electrical stimulation of mammalian central nervous system: a review. *Brain Res.* 98, 417, 1975.

58. McIntyre, C. C., and Grill, W. M. Selective microstimulation of central nervous system neurons. *Ann. Biomed. Eng.* 28, 219, 2000.

59. Drake, K. L., Wise, K. D., Farraye, J., Anderson, D. J., and BeMent, S. L. Performance of planar multisite microprobes in recording extracellular single-unit intracortical activity. *IEEE Trans Biomed. Eng.* 35, 719, 1988.

60. Cham, J. G., Branchaud, E. A., Nenadic, Z., Greger, B., Andersen, R. A., and Burdick, J. W. Semi-chronic motorized microdrive and control algorithm for autonomously isolating and maintaining optimal extracellular action potentials. *J. Neurophysiol.* 93, 570, 2005.

3 Thermal Considerations for the Design of an Implanted Cortical Brain–Machine Interface (BMI)

Patrick D. Wolf

CONTENTS

3.1 INTRODUCTION

Devices for the brain–machine interface (BMI) are no longer science fiction but are being tested in patients with severe paralysis and advanced neuromuscular diseases.

While the feasibility of controlling a device using brain-derived neural signals has been demonstrated, the first practical system for acquiring and processing these signals has yet to appear. This is in part because the implanted device technology needed to make a clinical BMI is substantially different in many ways from existing implanted devices. As described below, the BMI will use significantly more power than other implants. More power leads to the necessity of dissipating more heat inside the body.

This chapter discusses tissue heating by the BMI components, one of the engineering challenges that need to be solved in order to create a functional BMI. A functional overview of a cortical BMI is given, with particular attention paid to the sources of heat, followed by a section reviewing quantitative methods to examine these sources of heat. A review of the literature characterizing tissue response to heating follows, and finally recommendations are made regarding BMI system design based on considerations of heat dissipation.

3.2 OVERVIEW OF THE SOURCES OF HEAT IN A CORTEX BMI (cBMI)

A BMI is a device that uses brain-derived neural signals to control the operation of an external device [1–5]. A BMI can use external electroencephalogram (EEG)-derived signals [5], epicortically derived signals from electrodes placed on the surface of the brain (EcBMI) [6,7], and cortically derived signals from electrodes placed in the cortex (cBMI) [8–11]. Each of these technologies have advantages and disadvantages to both their implementation and their capabilities. Generally, the cBMI is the most invasive and provides the greatest potential in terms of function. This chapter deals specifically with the technical challenge of thermal management associated with implementing a totally implanted cBMI. In the context of this discussion, the term *totally implanted* carries the meaning of "no transcutaneous electrical conduit." This is in direct contrast to the transcutaneous connector used in the only clinical trials of a cBMI to date [12]. By definition and necessity, portions of a cBMI must be implanted in the brain and portions (the machine) must be outside of the body. The topic here is the implanted component, including the means to power it and for it to communicate with the nonimplanted component.

A complete cBMI would include devices to extract information from the brain and to input information to the brain. The general method for providing input is to stimulate neurons directly with either electrical or chemical stimulation. Stimulation has been implicated as a direct source of brain heating [13], but this means of heating will not be discussed here. Most specifically, this chapter will deal with the thermal problems encountered when designing an implanted system to extract and transmit information from individual cortical neurons.

Figure 3.1 is a block diagram of the implanted components of a cBMI. The electrodes transduce the signal from the brain to the amplifier array, which amplifies and filters the signals. The signal processing block processes and prepares the data for telemetry. The transcutaneous energy transfer system (TETS) receives power across the skin and conditions it for use inside the device. Each of these components will be discussed below.

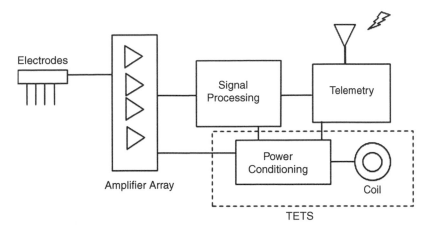

FIGURE 3.1 Block diagram of the implanted components of a cortical BMI. Electrodes acquire the voltage signal from the neural tissue. The amplifier array amplifies and filters the signal, and the signal processing hardware performs A/D conversion and implements data reduction schemes to prepare the data for telemetry. The telemetry component transmits the data to an external receiver.

3.2.1 SENSING CORTICAL BRAIN ACTIVITY-AMPLIFIERS

Individual neurons in the brain relay information from their inputs (dendritic arbor) along their length (axons) to their outputs (synaptic terminals). The information is carried as electrical action potentials that are conducted as spikes from the inputs to the outputs. The spikes can be characterized as brief (1 ms) changes in the electrical potential across the cell membrane that are propagated by active processes in the cell. The active process consists of two phases, depolarization and repolarization. The result is a brief ~100-mV change in the membrane potential. The spatial potential gradient that results from the propagation of these potential changes in the membrane voltage cause currents to flow in the extracellular space. These currents in turn cause potential differences in the body that can be detected by measuring the voltage difference between two spatial locations.

Typically, voltage differences are measured using two metal electrodes inserted into the extracellular milieu. One electrode acts as a reference or ground potential and the second as the sensing electrode. Because the currents created by the depolarization and repolarization processes are small, the sensing electrode must be placed very close to the membrane to measure a signal of sufficient amplitude to detect reliably. A great deal of work has been done to describe the relationship between the extracellular potential generated by a neuron and the depolarization and repolarization processes in the membrane. The spatial extent of this extracellular potential is small, and the amplitude falls off the square of the distance to the active membrane. This relationship severely limits the sensing range of an electrode. It is generally thought the maximum distance that an electrode can be from the active membrane to measure a detectable spike is on the order of 80 to 120 μm [14].

The amount of extracellular current being produced at any one instant from a given neuron is proportional to the amount of membrane generating the action potential at that instant. The most simultaneously activating membrane is present when the action potential propagates through the cell body. Thus, the greatest extracellular potential differences are generated near the cell body. This suggests that the greatest extracellular potential difference that can be measured for a given activating neuron is immediately adjacent to its cell body [15].

Electrode composition and manufacturing is the subject of intense research at this time. The interface to the extracellular milieu is generally metal, but carbon and conductive polymers are also being investigated. This work is discussed elsewhere in this book (Chapters 2, 6, and 7).

The electrode does not measure the electrical currents from one neuron but from all neurons in its vicinity [14]. The neural spikes from neurons close to an electrode have a peak amplitude of 100 to 500 µV. From zero to several hundred cell bodies can be close enough to the electrode to produce detectable spikes. This number depends on the types of neurons present and on the area of the brain from which recordings are made. In the motor cortex this number averages very near 2 neurons per electrode, with the range from 0 to 6 [16].

Neurons at greater distances produce a background spiking that also appears in the signal. This background firing results in a noise component of the signal that is generally tens of microvolts in amplitude. The resulting recorded spike signals have a signal to noise ratio (SNR) between 1 and 10. This signal to noise ratio is problematic in terms of extracting information as described below. Figure 3.2 shows an

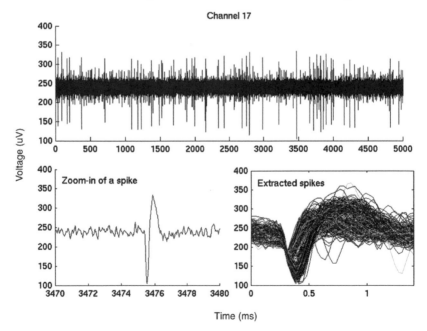

FIGURE 3.2 An example of neural signals recorded from the motor cortex of a monkey. The same signals are shown at three different time scales, with the final panel showing the spikes overlaid with a common detection point. **(See color insert following page 110.)**

example of signals recorded from a microwire in the motor cortex of a monkey. The signals are shown at several timescales for better visualization.

To be processed, the neural signals must be amplified and filtered. The number of neural signals that are needed to implement useful control of a device is the subject of some debate. It has been shown that the ability to guide a robotic arm to a limited number of targets in 2 and 3 dimensions can be implemented with tens of neurons [10,17]. Useful linear control in 2 and 3 dimensions has been shown using several hundred neurons [11]. Whichever type of system is created now, it is extremely likely that more sophisticated systems in the future will utilize hundreds to thousands of neural inputs. Thus, the power required to process the signal from an individual electrode must be small, as the multiplicative factor, the number of channels, is likely to be extremely large. For example, if a (yet unbuilt) neural amplifier can amplify and filter the neural signal from a single electrode using 1 µW of power, then just the amplification stage of processing for 1000 channels will consume 1 mW of power.

There have been numerous examples of integrated neural amplifiers developed over the past 5 years [18–24]. The amount of power required to amplify these signals is fundamentally related to the noise level in the first stage of the amplifier; more current in the transistors in this stage results in less noise [24]. Because the neural signals are small, the amount of noise that can be tolerated is small (<5 µV RMS). New designs that limit the supply voltage may reduce the power requirements by a small multiplicative factor, but traditional MOS technology may be reaching the limits of design improvements for lowering the power required for adequate amplification of neural signals.

3.2.2 SPIKE ACQUISITION AND PROCESSING

Most signal processing is performed in the digital domain, and this requires an analog to digital (A/D) converter to convert the waveforms to a series of digital values. Because the spike waveforms are short quick transients, sampling must be fairly high to ensure fidelity. Neural waveforms are usually sampled between 25 and 40 kHz for cBMI applications. As discussed below, continuous A/D conversion of each channel may not be necessary for a functional cBMI. However, if the research systems from the groups mentioned above, which have been used to demonstrate cBMI feasibility, are replicated in their functionality, the waveform from each electrode would be digitized, transmitted, and processed.

Understanding how information is transferred from the senses to the brain, from point to point within the brain, and from the brain to the motor system is the subject of the entire field of neurobiology. Generally, the information is thought to be relayed in the temporal patterns of the neural spikes from a single neuron.

As discussed above, spikes from multiple neurons, or units, are sensed by a single electrode. If the spikes are of sufficient amplitude, they can be discriminated in a process known as spike sorting. Spike sorting separates the multiunit neural signal into spike trains from individual neurons and a background spiking [25]. This background spiking can be further decomposed into detectable but nonsortable units and nondetectable noise spikes.

It has been shown that the information contained in sorted spikes can be used to predict arm movements and to control a robotic device [8,10]. It has also been shown that similar but less precise control can be attained using the detectable but unsorted spikes [8].

Spike sorting in the cBMI research environment is performed in a two-step process involving a learning phase, in which individual waveforms are defined in terms of their features, and a real time phase when the features are used to identify the spikes as they are generated. In an implanted cBMI device design published recently, the trend has been to ignore spike sorting and to use the detected multiunit signal as the information-carrying attribute [26,27]. The reason for this is the system complexity and additional power that must be added to the system to implement the sorting process.

Some recent studies have shown that the value of spike sorting in maintaining the information in the individual spike trains is limited by the difference in the information content of the signals and by the errors produced during the sorting process [28–30]. If decomposition of an electrode's multiunit signal into its individual neuronal components is found necessary for adequate control of a BMI device, then spike sorting will have to be performed either in the implanted system or in the external device. If it is performed in the implanted system, it will increase the signal-processing power consumed per channel but reduce the telemetry burden. If it is performed outside of the body, then enough information about each spike would need to be telemetered out of the body to allow sorting. This substantially increases the telemetric burden and increases the power consumption of the implanted system.

We have implemented a digital integrated circuit (IC) in the form of a field-programmable gate array (FPGA) that detects spikes and extracts the spike waveform segments for telemetry [29]. This system performs spike sorting outside the body using the extracted spike segments. A 16–20 to 1 data reduction is realized by transmitting only the detected spike waveform segments. The FPGA operates in real time on 96 channels of data and consumes almost 100 mW of power. We estimate that the power for this function could be reduced by a factor of 10 by implementing the functions in a custom-integrated circuit. Thus, we find that almost 0.1 mW per channel is necessary in the implant to perform spike detection, extraction, and sorting in this way.

3.2.3 Telemetry

The information carried in the monitored neuronal firings must be transmitted to a component outside the body to affect control of a machine. Radio telemetry of information requires power to generate the electromagnetic signal. For omnidirectional antennas, the power required goes up as the square of the distance between the transmitting antenna and the receiver. The distance to the receiver in a cBMI system would vary depending on the system design. If the receiving antenna is located on the body, as close as possible to the implanted antenna, the distance is on the order of 1 to 2 cm. If the receiver is located off the body, a telemetry distance of a few meters may be required. Because the transmitting antenna is located inside the body, a great deal of the transmission energy is deposited in the tissue surrounding the antenna. This energy loss attenuates the radio frequency (RF) signal substantially and heats the tissue directly. This heating is discussed below and must be considered in the system design.

The power required to transmit information is proportional to the bandwidth of the information signal [31]. Thus, the minimum bandwidth signal should be transmitted to minimize the power consumption in the telemetry component. If the implanted system digitizes and transmits the complete spike waveform, the transmission bandwidth required is substantial. For 1000 waveforms digitized to 10 bits resolution at a sampling rate of 25 kHZ each, the required bandwidth would approach 250 MHz. Such a system is not possible with the technology available for implants today. If spike sorting were implemented inside the body, yielding 2000 sorted neurons, the resulting spike streams could be transmitted out as individual bits for each channel with 1 mS resolution. This system would require a bandwidth approaching 2 MHz. This system is practical in terms of radio telemetry, but the internal power required to perform the sorting would increase the total power consumption of the implanted system. The attractiveness of using the multiunit signal as discussed above can be seen here. By simply detecting multiunit spikes and transmitting their occurrence at 1 mS resolution, the 1000-channel telemetry bandwidth can be reduced to about 1 MHz.

Many factors influence the efficiency of transmission in a radiotelemetry link. The transmitting and receiving antenna, the distance between the source and the receiver, the amount of power in the transmission, and the sensitivity of the receiver are all important determinants of the quality of the telemetry. These factors all impact the power requirements and the system design substantially. For example, we found that a commercial transceiver (RFM TR1100, RFM Inc., Dallas, Texas) consuming 36 mW (peak transmitted power 1 mW) and transmitting from a ½ wavelength dipole antenna located 2 cm below the skin can transmit 1 MBit of data per second a distance of 2 m. Harrison et al. reported a 0.33 MBit/second transmitter operating at 433 MHz that could transmit a distance of 13 cm using approximately 6.75 mW [26]. The transmitting antenna in this system was the 470-μm square inductor used in the oscillator circuit.

3.2.4 Powering the Implant

A typical pacemaker battery would last approximately 5 months continuously drawing 1 mW. The components of the system discussed above all use substantially more power than this. In terms of electrical power, cBMIs are a very different kind of implanted system than the typical pacemaker, defibrillator, or neural stimulator. At a minimum, rechargeable batteries will be required or, more likely, continuous TETS. Even in a system using rechargeable cells, transcutaneous energy transfer would be required to charge the cells. Thus, all currently considered cBMI devices will require either continuous or transient transcutaneous power delivery. [20,26,32,33]

Transcutaneous power delivery is implemented by magnetically coupling an external and an implanted coil creating a tissue core transformer. Figure 3.3 shows the components of our group's TETS system. The outside and implanted coils are shown, as well as the implanted electronics. A substantial amount of power can be transferred into the body using this technique [34].

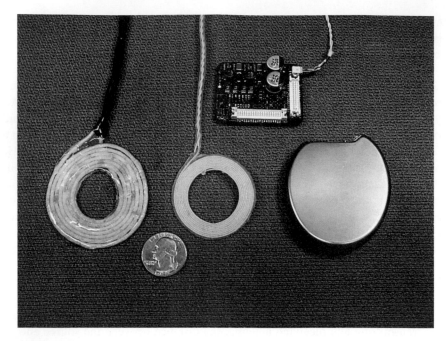

FIGURE 3.3 The TETS system used by our group to provide 1.6 W of power to an implanted device. The larger external coil excites the small implanted coil through the skin. The induced AC current in the inner coil is converted to DC voltages by the circuits shown on the PC board. This board also holds the signal processing and telemetry components of the cBMI system.

The internal components of a TETS system must rectify the AC waveforms required for magnetic coupling into the DC current required for operating the internal circuits. In addition, voltage regulation is required to actively match the transmitted power to the power consumed by the circuits. These functions cannot be performed with 100% efficiency but result in approximately a 10 to 20% power loss. If 100 mW of regulated power is required by the cBMI circuits, then approximately 110 to 120 mW would be dissipated within the body because of inefficiencies in the power circuitry.

The coupling in a tissue core transformer is relatively weak, making the efficiency of the link low. We have found that an external coil that is 6 cm in diameter coupled to an implanted coil 4 cm in diameter separated by a distance of 1 cm has an overall power efficiency of approximately 25%. To obtain 1.6 W of regulated power inside the body, approximately 6.4 W of DC power must be drawn from the external source.

3.3 HEAT GENERATED IN A cBMI IMPLANT

A critical issue that must be considered when discussing the power required for an implant is that of heat dissipation. Any electrically powered implanted device dissipates its electrical power as heat or radiates it as electromagnetic radiation. For the heat to be dissipated into the body, the temperature of the implant must be above that of the surrounding tissue. The magnitude of the temperature difference is what is critically important to the safety of the device. In a pacemaker, consuming microwatts of power, the temperature increase is almost negligible. In a cBMI consuming hundreds of milliwatts, using a transcutaneous energy delivery system and RF telemetry, the magnitude of the temperature increase and its effects will depend on many factors. Below, an attempt is made to identify the factors important to determining the temperature increase and to examine the effects a temperature increase may have on body tissue.

3.3.1 HEAT DISSIPATION IN THE BODY

The Pennes bioheat equation was developed in 1948 by Harry H. Pennes and has been the standard for describing heating in the body since that time [35]. The general form of the equation is:

$$\rho c \frac{\partial T}{\partial t} = \nabla \bullet (k \nabla T) + Q_v - Q_p + Q_m \tag{3.1}$$

where
ρ = tissue density (kg/m^3)
c = tissue specific heat (J/kg/$^\circ$C)
T = temperature ($^\circ$C)
t = time (s)
k = tissue thermal conductivity (W/m/$^\circ$C)
Q_v = volumetric heat source (W/ m^3)
Q_p = heat loss due to perfusion (W/ m^3)
Q_m = metabolic heat generation (W/ m^3)

The thermal properties of tissues are summarized in several places, but the parameters of tissues in the head are conveniently enumerated in a paper by Lazzi [36].

The left side of this equation represents the change in tissue temperature over time. This term will be zero when considering a time average of the whole body. In this state, the terms on the right side of the equation balance, representing the homeostatic condition of a constant (in time) normal body temperature. This term will also be zero when considering the long-term steady state response to a localized constant heat source, for instance, as would exist in the tissue adjacent to a chronic cBMI implant. If the heat flux generated by the device is constant, it will result in a steady state balance of the terms on the right side. The left-side term will be nonzero when the heat generated by the operation of the implant is time varying, for example, during battery recharging, or when the terms on the right side are time varying, such as when perfusion or metabolic heat generation increases or decreases transiently in the vicinity of the implant.

The first term on the right side of Equation (3.1) describes the diffusion of heat into the cooler tissue surrounding the source. The diffusion process is driven by the temperature gradient, increasing with temperature difference. The clearance of heat from a specific location in the body is driven by this term. Heat diffuses through the tissue into the cooler circulating blood. The blood circulates, and as it reaches cooler areas of the body, generally the skin and lungs, the heat diffuses out of the blood into the local tissues and then out of the body. Heat removal from the lungs and from the skin is governed by equations dominated by convection and is strongly influenced by environmental conditions. Our discussion here is limited to heat generation and clearance at the source within the body.

The term Q_v represents external volumetric heat sources. In the context of a cBMI, these sources include the heat generated within any implanted device, the heat deposited within the tissue by the electromagnetic fields associated with transcutaneous energy transfer and transcutaneous telemetry. An electronic device dissipating a certain amount of electrical power can be modeled as a constant volumetric heat source (W/m³), or the device volume can be considered outside the domain of the solution and the boundary of the device modeled with a boundary condition of either a constant temperature or constant heat flux (W/m²).

The Q_p term is the heat lost because of perfusion and is, to some extent, an averaged term. The term is tissue specific and varies broadly with the range of tissue perfusion found within the body. In general, the body clears excess heat in specific tissues by increasing Q_p. Thus, evaluation of this term is critical to understanding how the temperature changes in response to external heat sources. Because the rate at which heat is carried away by perfusion is dependent on the temperature difference between the blood and the surrounding tissue, the Q_p term can be written as $Q_p = B(T - T_B)$, where B is a tissue-specific capillary blood perfusion coefficient (J/m³ s °C), T is the temperature as a function of position and time, and T_B is the blood temperature. The range of values for B varies widely by tissue and experimental condition. For example, B for fat is 1,700 J/m³ s °C, while for brain it is 40,000 J/m³ s °C.

Q_m is the metabolic heat generated by the tissue. Metabolic heat generation is a function of metabolic activity and is also tissue specific. Heat generation due to metabolic activity of a tissue can vary substantially with function, over time, and with temperature. In the brain, for instance, heat from the metabolic activity of individual cells can vary from 0.5 to 4.0 nW depending on location and level of activity. This is a substantial range and is 300 to 2500 times what the average body cell produces (1.6 pW) [37].

The Pennes equation is extremely useful when discussing the qualitative effects of heating and for understanding the general concepts of tissue heating, but the equation is difficult to use to make quantitative predictions of temperature changes. Quantitative analysis requires this equation be solved in a well-defined domain with accurate boundary conditions. The domain of the body is particularly difficult in this regard because of the many tissue types, the irregular geometry, and the range of boundary conditions. In general, each tissue type has unique physical properties:

density (ρ), specific heat (c), and thermal conductivity (k). As discussed below, each tissue type also has a unique time-dependent heat flux provided by perfusion and a time-dependent metabolic heat generation term.

As electromagnetic waves propagate through tissue, energy is absorbed. The specific absorption rate (SAR) is a measure of the volume heating caused by the electromagnetic field in tissue.

$$SAR = \frac{\sigma |E|^2}{\rho} \ (W/kg) \tag{3.2}$$

where $|E|$ is the magnitude of the complex value E representing the electric field.

This relatively simple equation masks the difficulty in computing the value for the two sources of electromagnetic (EM) radiation discussed here. The difficulty is caused by the need for an accurate calculation of the E field in the tissue. The recommended SAR limit for tissue is 1.6 W/kg or 1.6 mW/g (average over 1 g) for radiation in the 3 kHz to 300 GHz spectrum [38].

For low-frequency TETS, where the system can be modeled as a transformer with a tissue core, three factors dominate heat generation in tissue. First, there is the heat generated by current flowing in the coils [39]. This can be calculated by multiplying the RMS current squared by the equivalent series resistance of the coil at the frequency of the link.

The second source of heat is the direct heating of the tissue caused by the E-field generated by the coil. The calculation of this E-field is complex but can be done by calculating the magnetic field (H field) using the current flowing in wire segments and then calculating the E field from the H field. Usually, the operating frequency of these devices is low (hundreds of kHz), leading to low absorption of this RF energy and negligible heating of the local tissue.

The third source of heating comes from the unintended induction of eddy currents in metal objects located in the magnetic field generated by the primary coil [40]. These currents can generally be minimized by appropriately separating metal objects from the coil in the body. However, the design of the cBMI system may make such separation impossible, as in the case of a completely self-contained unit that includes electrodes, amplifiers, signal processing, and power conversion circuitry in a single package. Proper selection of materials and careful consideration of geometry is required to minimize eddy currents in these systems. For low-frequency, large-coil systems, the dominant heating term is usually the coil current, assuming all potential eddy current sources have been minimized.

The second model for a TETS is to transmit power using radiation from an RF source located some distance from the body. In this mode, the E field is primarily responsible for the power transfer. When the source is some distance from the body, the RF waves can be considered as impinging on the body as plain waves. As plain waves strike the body surface, some of the waves are reflected and some transmitted. The reflection coefficient can be calculated based on the frequency and the material properties of the interface. While the reflection coefficient is important for determining the efficiency of the system, only the transmitted waves will cause heating of the tissue. Once inside the body, the waves are attenuated as they propagate through the tissue. The energy lost in this way is primarily dissipated as heat. The equations governing the heat generation in this case are shown below [41].

The power density as a function of depth within the tissue is:

$$W_d = We^{-2\alpha d} \; (W/m^2)$$

(3.3)

where W is the transmitted portion of the impinging wave (W/m²), d is distance (m) and α is the attenuation coefficient of the tissue and is defined as:

$$\alpha = \omega \sqrt{\frac{\mu\varepsilon}{2}} \left[\sqrt{1 + \left(\frac{\sigma}{\omega\varepsilon}\right)^2} - 1 \right]^{\frac{1}{2}} \; (Neper/m)$$

(3.4)

where $\omega = 2\pi f$, the frequency of the incident wave, and μ, ε, and σ are, respectively, the permeability, the relative permittivity, and the conductivity of the tissue [42].

The relationship between power density and the E field is:

$$W_d = \frac{1}{2} \mathrm{Re}\left(\frac{1}{\eta}\right) |E|^2$$

(3.5)

where η is the intrinsic impedance, defined as:

$$\eta = \frac{\sqrt{\mu/\varepsilon}}{\sqrt{1 - \dfrac{\sigma}{j\omega\varepsilon}}}$$

(3.6)

Thus, the SAR for this case can be calculated as:

$$SAR = \frac{2\sigma W e^{-2\alpha d}}{\rho \, \mathrm{Re}\left(\dfrac{1}{\eta}\right)}$$

(3.7)

Here, Re() denotes taking the real part of the complex variable.

In general, because of the frequency dependence of the electrical properties and because of the frequency dependence of Equation (3.4), the absorption is strongly frequency dependent. Frequencies above 1 GHz do not propagate far in body tissues. However, if frequencies below 100 MHz are used, the SAR from this source is manageable. It was shown in a modeling study by Ibrahim et al. [43] that the RF field for powering a device at 13.6 MHz contributed negligibly to the SAR and tissue heating in a multicompartment model of the head.

3.3.2   Tissue Heating Caused by Radio Telemetry

Radio waves impinge on our bodies and interact with the tissue continuously. How radio waves interact with tissue has become of great interest recently because of the proliferation of personal communication devices, in particular cell phones.

A significant body of literature exists characterizing the effects of cell phone RF energy on tissues [44–46]. The modeling and testing of the thermal effects of these devices is quite complicated and is made increasingly so by the proliferation of devices at different operating frequencies and with differing antenna configurations. It is very difficult to exactly predict the heating that will occur in tissue from radiated RF energy.

In an important paper in 1992, Kuster and Balzano [45] demonstrated that the induced SAR in the extreme near field of antennas for frequencies above 300 MHz was primarily related to the antenna current. The antenna currents generate magnetic fields (H fields) in the tissue that induce currents that heat the tissue. The SAR was related to H^2, and these fields fell off inversely with the square of the distance from the feedpoint of the antenna. These studies were done to model cell phone operation and did not specifically address implanted antennas.

For implanted telemetry systems, the antenna is placed in the tissue, and this leads to a more complex situation. In one of the few studies addressing this configuration, Kim and Rahmat-Samii [47] modeled transmitting antennas implanted in the head and thorax. In this study, the authors characterized the radiation patterns and SAR for several small antennas operating at 405 MHz. They modeled small microstrip antennas placed near the surface of the thorax and showed that the delivered power to the antenna should be limited to less than 9 mW to stay below the recommend SAR levels (1.6 mW/g) in the tissue surrounding the antenna. They also showed that the radiation efficiency for these configurations is extremely poor (0.16%), indicating the difficulty radiating RF energy out of the body. This is significant for heating because most of the energy lost is dissipated as heat in the tissue.

SAR deposited in the tissue from the telemetry source is likely a significant contribution to the total heat burden. Because this energy heats the tissue directly, the effect will be additive to heat diffusing into a given area. For instance, if the heat next to an implant is causing a 1°C temperature increase and a nearby antenna is made active, the RF radiation will cause an additional temperature increase.

3.3.3   Insulation

One might think that if an electronic device dissipating power were insulated, it would prevent temperature increases in the body. Insulation is valuable in this situation, as it prevents localized high heat sources from generating high local temperatures. In other words, insulation can be used to lower peaks and raise the valleys, but the average heat dissipation over time and space will remain the same.

If the electronic power dissipated within a fixed-volume device remains constant, then at steady state, the thermal flux from the device will also be constant. When a device is first turned on, the heat flux will be absorbed by the insulation, raising its temperature. In turn, the temperature of the electronics will increase and so on until a steady state temperature gradient is achieved between the electronics and the insulation and between the insulation and the body that causes the exact amount of heat generated by the implant to diffuse into the tissue. Thus, while insulation is useful for preventing localized temperature increases, in the steady state, it does not reduce the thermal burden presented to the body.

Many implanted devices are encased in metal. The effect of this encasement on heat dissipation is essentially the opposite of insulation, as metals have extremely high thermal conductivities. Interestingly, the thermal effect of the metal boundary is similar to that of insulation in the steady state. The high thermal conductivity of the can distributes the heat evenly over the surface, minimizing hot spots. In the steady state, however, the average thermal burden on the tissue is the same with or without a can.

3.3.4 BMI-RELATED THERMAL STUDIES

The heat generated in implants has been studied in some specific devices related to BMIs. There has been an extensive treatment of this subject by the group designing the retinal prosthesis [32,36,48]. This device is primarily a high-density–high-channel count stimulator designed to restore some vision by directly stimulating the retina or optic nerve. The device is designed to lie completely within the optical cavity and is powered by a TETS system operating from the front of the eye where the coil might be contained in a pair of glasses. Stimulation data is telemetered into the device.

In one study, Gosalia et al. [32] modeled a heater placed in the eye and compared the modeling results with experimental measurements of temperature near an acutely implanted thermal source. The thermal source was a $1.4 \times 1.4 \times 1$ mm probe. They found close agreement between the model's predicted temperature increase and the temperature increase measured in the eyes of dogs. They found that the temperature increase in the tissue related to the RF energy of the TETS system (10 MHz, 2A coil 20 mm in front of the eye) provided a negligible increase in the temperature of the eye when compared to the contribution of the direct heating by the implant. They also found in simulations that the vitreous cavity provided a high thermal conductivity heat sink for the implant with high temperatures at the implant surface but substantially lower temperatures in other structures in the eye. A power dissipation of 12.4 mW from the implant caused only a 0.82°C increase in the temperature at the device surface.

It is interesting to note that when the total size of the implant is calculated, including the insulation layer, it is found to have a surface area of 0.8 cm². When the device provided a heat flux of 12.4 mW/0.8cm² = 15.5 mW/cm², the temperature increases were less than a degree. When the power was increased to represent 62 mW/cm², the temperature increases reached as high as 3°C. The Davies group (below) found that 40 mW/cm² was the limit to prevent significant temperature increases in muscle but that 60 mW/cm² caused similar increases to those seen here. It seems this heat flux guideline may apply more generally in the body.

3.4   EFFECT OF HEAT ON TISSUE

The effect heat has on tissue is indeed complex. Consider, for instance, the therapeutic application of hyperthermia. All of us are aware of the beneficial effects of heat on injured and aching muscles and joints. This is accomplished with a degree or two temperature increase over the affected area for tens of minutes. If the temperature is increased a few degrees (42°C) and the duration of heating increased (hundreds of minutes), these beneficial effects turn deadly to cells. This is the phenomenon used in hyperthermia treatment of certain cancers. Increase the temperature still further and apply the heat for only a minute, and it can be similarly deadly for tissue. This technique, called ablation, is commonly used to kill arrythmogenic tissue in the heart [49]. Thus, the effect of a temperature increase on tissue is primarily related to the magnitude of the temperature increase but is also critically related to the length of time the temperature is elevated and the tissue to which the heat is applied.

For a CBMI, we are interested in understanding the effects of a chronic temperature increase in the range of 1 to 2°C. This number is the limit recommended by the American Association of Medical Instrumentation (AAMI) for implanted medical devices. The amount of power that can be dissipated by the body and still remain within this limit is a question of great importance. As can be seen from the discussion of the bioheat equation above, the ability of tissues to dissipate heat is dependent on many factors but is primarily related to tissue perfusion.

The reaction of the body to relatively high heat sources was the subject of a series of studies related to powering the implanted artificial heart [50–52]. In these studies, constant power density heat sources were implanted in the latissimus dorsi muscle and adjacent to the left lung in calves. The constant power densities tested were 40, 60, and 80 mW/cm^2. The sources were implanted for 2, 4, and 7 weeks to examine the changes in temperature in the tissue surrounding the implants over time. Temperatures were monitored *in vivo* at locations a few millimeters from the source. The use of a power density (W/cm^2) in these experiments can be equated to implanting a device with a surface area of X cm^2 dissipating a total of Y Watts within the device. For instance, a device dissipating 1 Watt with a total surface area of 25 cm^2 has a power density of 40 mW/cm^2.

In this study, temperatures in the tissue adjacent to the heat sources decreased over time. For example, the implant surface temperatures of the 60-mW source were 4.5, 3.4, and 1.8°C above normal at 2, 4, and 7 weeks, respectively. Histopathological results showed that the body responded to the presence of the heat source by forming a fibrous capsule with increased capillary density around the implant. The net effect of the angiogenesis was to decrease the temperature around the implant over time by increasing the heat carried away by perfusion [50–52].

The initial temperature increase for the 40 mW/cm^2 source was 1.8°C, suggesting this as an upper limit to the power density that can be handled by these tissues while staying under the 2°C recommended limit.

A histopathological study of the muscle tissue response to the 80 mW/cm^2 source versus control unheated implants found a consistent line of muscle tissue necrosis at 42.1°C (38.6°C is normal body temperature for calves) at 2 and 4 weeks. The front of angiogenesis ended at a constant temperature line of about 41.7°C, suggesting endothelial cell temperature limits in this range. Overall, capillary densities were 4 times greater in the heated versus unheated implant controls [50].

3.4.1   Brain Tissue Response to Heating

The studies by Davies et al. [50–52] demonstrated that the 2°C temperature limit recommended by the AAMI causes a physiologic response by muscle tissue that is well tolerated at 7 weeks. There is no similar work that shows the response of neural tissue to these types of thermal loads.

The brain has high resting blood flow rates; it receives 20% of the cardiac output despite representing only about 2% of the body's mass. This indicates that there is an extremely high basil metabolic need in the brain, the heat from which is cleared by the high blood flow. Kiyatkin demonstrated by making simultaneous brain and core blood temperature measurements in rats that the net flow of heat is from the brain under normal conditions [53]. The high metabolic rate in the brain is fed by the high blood perfusion rate, which also clears the high metabolic heat load. In effect, the balance discussed earlier in the context of the bioheat equation is maintained at a substantially higher level in the brain than in other tissues.

The increased capacity for thermal clearance would seem to indicate that the thermal sources placed in the brain or on the brain surface might be well tolerated. No studies were found that characterize the response of brain tissue adjacent to a thermal source.

Increased perfusion and hence heat clearance does not address the issue of the sensitivity of the brain tissue to increased temperature. It is generally agreed that central nervous system (CNS) tissue is one of the most sensitive to heat [54,55]. Adverse effects to the CNS have been observed in animals at relatively low temperature increases [54]. Blood–brain barrier breakdown was shown after 60 minutes at 42°C, and some neuronal death was observed (6%) after 60 minutes of heating at 40.5°C. A review of the animal studies of CNS damage following regional exposure of the brain and spinal chord established a maximum tolerable heat dose of 42.0 to 42.5 °C for 40 to 60 minutes or 43°C for 10 to 30 minutes [55]. This dose was established based on injury endpoints, not just response endpoints. Induction of heat shock proteins in neural support cells was seen at lower temperatures and was responsible for a cascade of effects that induced heat tolerance. All of the studies reviewed considered time durations of heat application of 90 minutes or less. Studies of chronic exposure to low temperature sources in the brain are needed to determine if there is a negative CNS response to these loads.

Very little data exist about long-term (thousands of minutes) application of very low (2°C) temperature increases on *in vivo* human or animal tissue. The cancer hyperthermia community is interested in this question, as they use heat to kill tumor cells and hope to do as little collateral damage as possible. Heat affects all tissues in all species, but the time and temperature profiles required to reach specific endpoints varies. Experiments have been performed on cells *in vitro* and on tissues *in vivo,* and techniques have been developed to try to unify the data and extend the time and temperature range of its validity.

It is possible to create an Arrhenius plot of the heat of cell inactivation by studying the number of cell deaths occurring *in vitro* as a function of time and temperature. For a review, see the articles of Dewey [56] and Dewhirst [57]. Extrapolation of these curves to the long-duration–low-temperature increase region could give some insight into the effect of the long-term hyperthermia related to an implant.

Arrhenius plots of cell inactivation show a significant change in slope or "breakpoint" of around 43°C. The inactivation energy above the breakpoint is in the range of 120 to 150 kcal/mole. This value is consistent with the heat of inactivation of proteins and enzymes. Below the breakpoint, the slope is greater (lower rate of inactivation), which is thought to indicate the development of thermo-tolerance during the heating process. While these curves vary in their absolute temperature values across species, the slope of the curves tend to be relatively similar across tissue types [57].

It is also possible to create Arrhenius curves for *in vivo* experiments using the time to an isoeffect (cell necrosis, first- and second-degree burn, etc.) at a given temperature. These curves show a similar breakpoint at approximately 43°C.

To compare studies occurring at different temperatures and time durations, an empirical relationship was developed to normalize results to heating at 43°C [57]. The equation for thermal isoeffective dose is:

$$CEM\,43°C = tR^{(43-T)} \tag{3.8}$$

where $CEM43°C$ is the cumulative number of equivalent minutes at 43°C required to reach the effect, and t is the time in minutes of the application of temperature T to the target tissue. R in this equation is related to the slope of the Arrhenius curve, either above or below the break point, and is the number of additional minutes it takes to reach the isoeffect for a 1°C temperature change.

The value of R is critical, and it is thought to lie in the neighborhood of 0.43 for human cells *in vitro* above the breakpoint and 0.23 below it. The location of the breakpoint seems to vary with species. The limited amount of data for human cells appears to show it to be 43.5°C. For human skin, which is the only tissue with extensive *in vivo* data available, the values for R are approximately 0.72 above the breakpoint and 0.13 below the breakpoint. The value below the breakpoint shows the extremely high tolerance of skin to temperatures below 43.5°C.

As an example of how this relationship can be used, if the R value is 0.2 and the breakpoint is 43.5°C for a particular tissue type, a plot can be made of the time–temperature relationship if one piece of data regarding an effect at a given temperature and duration is known. If an experiment is performed that demonstrates that 6% of this tissue's cells die when exposed to 40.5°C for 60 minutes, then the $CEM43°C$ can be calculated to be 1.07 minutes. This predicts that 6% of the cells would die if the tissue were exposed to 43°C for one minute. The $CEM43°C$ can then be used to predict how long it would take 6% of the cells to die at 39.5°C. Solving Equation (3.6) for t and plugging in these values results in a predicted duration of 300 minutes. Thus, this analysis predicts that it would take 5 hours at 39.5°C to reach the same effect. Unfortunately, the equation also predicts that 6% of the cells would die after 11 days at 37°C, so there is some difficulty using this relationship to predict effects at the very low temperature elevations associated with an implanted device. While these empirical equations are useful for unifying experimental data and extending its range of application, extrapolation must be done with care. Additional data on the parameters of these curves, in particular for animal and human brain tissue, is needed to improve these predictions.

There is potentially a very interesting question of cause and effect in mild brain hyperthermia. Regional temperatures in the brain vary over a substantial range. Kiyatkin showed by direct measurement in rats that normal brain temperature ranges over several degrees and demonstrated repeatable regional temperature increases of up to 1.8°C in response to specific behavioral stimulus [53]. Even the basal temperature within the brain seems to follow a dorso-ventral gradient of approximately 1°C. While there has been some controversy over the cause of this gradient, it appears that the difference in the basal activity of these regions may be the cause. Thus, normal brain temperature seems to vary regionally by a few degrees under normal conditions and increase regionally with local brain activity levels [37].

It could be that regional increases in temperature are used by the brain to regulate body temperature or to modify or enhance function or structure in the affected region. It is known, for instance, that dopamine uptake doubles with a 3°C temperature increase [58]. Similar temperature increases also affect the function of specific ionic channels. Regional hyperthermia may be used by the brain to increase the rate of structural changes important to learning or behavior. It is unclear from these studies if a chronic regional change in temperature, on the order of 1 or 2°C, as might be induced by a cBMI implanted in contact with the cortical surface, would affect changes in brain function that would be reflected behaviorally or psychologically. It has been shown, for instance, that regional heating of the hypothalamus in rats affects sleep and EEG patterns [59] and also has a significant effect on thermal regulation of the body. [60,61] Similar studies on local heating in the cortex need to be performed to quantify what thermal load can be tolerated before there are measurable affects on behavior.

The brain and CNS are sensitive to heating. Very few, if any, studies have looked at extremely long periods of low-temperature heating in the brain. Heat shock proteins that provide a protective response to heating are active at fairly low temperatures and begin a cascade of CNS effects [62]. Behavior also seems to cause regional increases in temperature in the brain. Further research is required to quantify the effects of low-temperature regional hyperthermia in the brain.

3.5 RECOMMENDATIONS

In this chapter, we have attempted to identify thermal load as an important consideration in the design of implanted cortical BMIs. The number of recording sites, the sophistication of the likely signal processing chain, and the need for telemetry and transcutaneous energy delivery will all push the power consumption of an implanted device toward the point at which the thermal burden for the body will be a critical factor. Thermal considerations lead to some recommendations regarding implant design that may help mitigate thermal effects. We make these suggestions recognizing they may be flawed, given the paucity of data regarding the physiologic response to heat of the myriad of tissues in the body.

3.5.1 RECOMMENDATION 1: DISTRIBUTE THE THERMAL LOAD

Interpretation of the studies by Davies et al. [51] and consideration of the bioheat equation show that local temperature increases can be minimized by distributing the thermal load in the body over space. The thermal capacity of a given tissue is dominated by its perfusion. Regional increases in perfusion seem to accompany regional changes in temperature. This effect is seen acutely with vasodilatation and chronically with angiogenesis and an increase in capillary density. The numbers from Davies indicate that a surface heat source density of 40 mW/cm^2 is the upper limit for implants located in muscle and lung. Similar studies in the scalp, the skull bone, and the cortex are required to determine if similar loads are tolerated by these tissues. Nonetheless, distributing a fixed thermal burden over a greater surface area will reduce the temperature increases associated with the load. While there is tremendous pressure to minimize the size of neural implants, the difficulty of managing the thermal burden in a small package only increases as the package size decreases.

Figure 3.4 shows one way such a distributed system could be implemented. The amplifier arrays are located on the head near the electrodes for maximum SNR. The arrays are contained in three packages (for example) to further distribute the heat across the scalp. The data processing and telemetry are located in an implant site in the thorax. This is a substantial distance from the scalp, far enough for the two thermal loads to be considered independent.

The coil for the TETS is located a short distance from the electronics to separate these thermal sources. The transmitting antenna is located several centimeters from the implant so that the heat generated in the extreme near field of the antenna does not add to the thermal burden generated by the device.

3.5.2 RECOMMENDATION 2: USE A LOW TRANSMISSION FREQUENCY FOR TELEMETRY

The absorption coefficient (α) for RF signals in tissue is strongly frequency dependent. Tissue absorption of RF energy can be reduced by using a lower transmitting frequency. Typically, the higher the bandwidth requirements of a transmitter, the higher the carrier frequency. New high-bandwidth–low-carrier-frequency transmitters may need to be designed to accommodate this need [63].

3.5.3 RECOMMENDATION 3: SEPARATE POWER TRANSMISSION COILS FROM OTHER IMPLANTED METALLIC OBJECTS

The induction of eddy currents in an implanted metal object can be minimized by increasing the distance between the magnetic field and the metal objects.

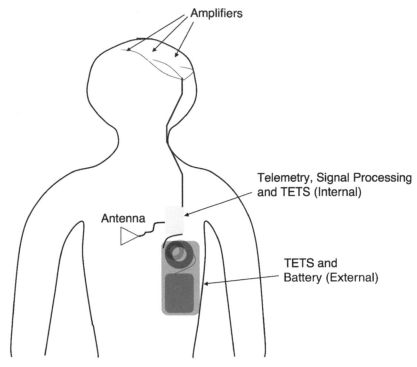

FIGURE 3.4   A schematic diagram of how the components of a cBMI system might be distributed to minimize the likelihood of thermal overload in any one location. Amplifiers are placed on the head near the electrodes; the remainder of the system is located in the pectoral muscle and is implanted similar to a pacemaker. The TETS coil is offset from the device to limit eddy currents and to further distribute the heat. The antenna is similarly offset to prevent the additional heating associated with the SAR of the RF radiation. The red component is the worn portion of the TETS system positioned so the coils overlap.

3.5.4   RECOMMENDATION 4: MINIMIZE THE POWER CONSUMPTION IN DEVICES IN CLOSE PROXIMITY TO THE BRAIN

Regional brain temperature in rats increases by up to 1.8 degrees for hours at a time in response to increased metabolic demand. A thermal burden of one or two degrees on top of this increase could put the total temperature increase in the range where a thermal response is provoked or injury may occur. These facts indicate that if you are going to increase the thermal burden in the body, it is probably best not to do it in the brain.

As described above, the acquisition and transmission of neural signals from the cortex is a complex task involving several processing steps. Neural amplifiers need to be placed in close proximity to electrodes to maintain an adequate SNR. However, the amplifiers can be separated from the electrodes by a short distance above the bone, providing a critical insulation barrier and limiting the direct heating of brain tissue.

3.6 CONCLUSION

The acquisition, processing, and telemetry of hundreds or thousands of neural signals requires significant power. The heat from the implanted system implementing these functions must be dissipated adequately to prevent an adverse tissue response to the implanted device.

The indications are that the limits of a 2°C temperature increase, of 40 mW/cm² heat flux, and of 1.6 mW/g of SAR are valid for most tissues in the body. The apparent sensitivity of CNS tissue to small temperature increases, the indications that regional metabolic heat generation in the brain may exceed the clearance capacity for unknown reasons, and the fact that there is little research regarding very long-term hypothermia in neural tissue suggests that using the brain tissue as a sink for device-related heat may be questionable.

REFERENCES

1. Donoghue, J. P. Connecting cortex to machines: recent advances in brain interfaces. *Nat. Neurosci.* 5(Suppl), 1085–88, 2002.
2. Nicolelis, M. A. Brain-machine interfaces to restore motor function and probe neural circuits. *Nat. Rev. Neurosci.* 4(5), 417–22, 2003.
3. Nicolelis, M. A. and Chapin, J. K. Controlling robots with the mind. *Sci. Am.* 287(4), 46–53, 2002.
4. Schwartz, A. B. Cortical neural prosthetics. *Annu. Rev. Neurosci.* 27, 487–507, 2004.
5. Wolpaw, J. R., Birbaumer, N., McFarland, D. J., Pfurtscheller, G., and Vaughan, T. M. Brain-computer interfaces for communication and control. *Clin. Neurophysiol.* 113(6), 767–791, 2002.
6. Leuthardt, E. C., Miller, K. J., Schalk, G., Rao, R. P., and Ojemann, J. G. Electrocorticography-based brain computer interface—The Seattle experience. *IEEE Trans. Neural. Syst. Rehabil. Eng.* 14(2), 194–198, 2006.
7. Leuthardt, E. C., Schalk, G., Wolpaw, J. R., Ojemann, J. G., and Moran, D. W. A brain–computer interface using electrocorticographic signals in humans. *J. Neural Eng.* 1(2), 63–71, 2004.
8. Carmena, J. M., Lebedev, M. A., Crist, R. E., O'Doherty, J. E., Santucci, D. M., Dimitrov, D. F., Patil, P. G., Henriquez, C. S., and Nicolelis, M. A. Learning to control a brain-machine interface for reaching and grasping by primates, *PLoS Biol.* 1(2), E42, 2003.
9. Chapin, J. K., Moxon, K. A., Markowitz, R. S., and Nicolelis, M. A. Real-time control of a robot arm using simultaneously recorded neurons in the motor cortex. *Nat. Neurosci.* 2(7), 664–70, 1999.
10. Taylor, D. M., Tillery, S. I., and Schwartz, A. B. Direct cortical control of 3D neuroprosthetic devices. *Science* 296(5574), 1829–32, 2002.
11. Wessberg, J., Stambaugh, C. R., Kralik, J. D., Beck, P. D., Laubach, M., Chapin, J. K., Kim, J., Biggs, S. J., Srinivasan, M. A., and Nicolelis, M. A. Real-time prediction of hand trajectory by ensembles of cortical neurons in primates. *Nature* 408(6810), 361–65, 2000.
12. Donoghue, J. P., Nurmikko, A., Black, M., and Hochberg, L. R. Assistive technology and robotic control using motor cortex ensemble-based neural interface systems in humans with tetraplegia, *J. Physiol.* 579(3), 603, 2007.

13. Elwassif, M. M., Kong, Q., Vazquez, M., and Bikson, M. Bio-heat transfer model of deep brain stimulation-induced temperature changes. *J. Neural Eng.* 3, 306–15, 2006.

14. Buzsaki, G. Large-scale recording of neuronal ensembles. *Nat. Neurosci.* 7(5), 446–51, 2004.

15. Holt, G. R. and Koch, C. Electrical interactions via the extracellular potential near cell bodies. *J. Comput. Neurosci.* 6(2), 169–84, 1999.

16. Nicolelis, M. A., Ghazanfar, A. A., Faggin, B. M., Votaw, S., and Oliveira, L. M. Reconstructing the engram: simultaneous, multisite, many single neuron recordings. *Neuron* 18(4), 529–37, 1997.

17. Serruya, M. D., Hatsopoulos, N. G., Paninski, L., Fellows, M. R., and Donoghue, J. P. Instant neural control of a movement signal. *Nature* 416(6877), 141–42, 2002.

18. Harrison, R. R. and Cameron, C. A low-power low-noise CMOS amplifier for neural recording applications. *IEEE J. Solid-State Circuits* 38(6), 958–65, 2003.

19. Neihart, N. M. and Harrison, R. R. Micropower circuits for bidirectional wireless telemetry in neural recording applications. *IEEE Trans. Biomed. Eng.* 52 (11), 1950–59, 2005.

20. Olsson, R. H., 3rd, Buhl, D. L., Sirota, A. M., Buzsaki, G., and Wise, K. D. Band-tunable and multiplexed integrated circuits for simultaneous recording and stimulation with microelectrode arrays. *IEEE Trans. Biomed. Eng.* 52(7), 1303–11, 2005.

21. Song, Y. K., Patterson, W. R., Bull, C. W., Beals, J., Hwang, N., Deangelis, A. P., Lay, C., McKay, J. L., Nurmikko, A. V., Fellows, M. R., Simeral, J. D., Donoghue, J. P., and Connors, B. W. Development of a chipscale integrated microelectrode/microelectronic device for brain implantable neuroengineering applications. *IEEE Trans. Neural Syst. Rehabil. Eng.* 13(2), 220–26, 2005.

22. Perelman, Y. and Ginosar, R. Analog front-end for multichannel neuronal recording system with spike and LFP separation. *J. Neurosci. Methods* 2005.

23. Obeid, I., Morizio, J. C., Moxon, K. A., Nicolelis, M. A. L., and Wolf, P. D. Two multichannel integrated circuits for neural recording and signal processing. *IEEE Trans. Biomed. Eng.* 50(2), 255–58, 2003.

24. Harrison, R. R. and Charles, C. A low-power low-noise CMOS amplifier for neural recording applications. *IEEE J. Solid-State Circuits* 38(6), 958–65, 2003.

25. Lewicki, M. S. A review of methods for spike sorting: the detection and classification of neural action potentials. *Network Computation Neural Syst.* 9(4), 53–78, 1998.

26. Harrison, R. R., Watkins, P. T., Kier, R. J., Lovejoy, R. O., Black, D. J., Greger, B., and Solzbacher, F. A low-power integrated circuit for a wireless 100-electrode neural recording system. *IEEE J. Solid-State Circuits* 42(1), 123–33, 2007.

27. Olsson, R. H. III and Wise, K. D. A three-dimensional neural recording microsystem with implantable data compression circuitry. *IEEE J. Solid-State Circuits* 40(12), 2796–804, 2005.

28. Won, D. S. and Wolf, P. D. A simulation study of information transmission by multiunit microelectrode recordings. *Network Computational Neural Syst.* 15, 29–44, 2004.

29. Rizk, M., Obeid, I., Callender, S. H., and Wolf, P. D. A single-chip processing and telemetry engine for an implantable 96-channel neural data acquisition system. *J. Neural Eng.* 4, 309–21, 2007.

30. Won, D. S., Chong, D. Y., and Wolf, P. D. Effects of spike sorting error on information content in multi-neuron recordings. *Conference Proceedings. First International IEEE EMBS Conference on Neural Engineering*, pp. 618–621. IEEE Press, Washington, DC, 2003.

31. Shannon, C. E. A mathematical theory of communication. *Bell Sys. Tech. J.* 27, 379–423, 1948.
32. Gosalia, K., Weiland, J., Humayun, M., and Lazzi, G. Thermal elevation in the human eye and head due to the operation of a retinal prosthesis. *IEEE Trans. Biomed. Eng.* 51(8), 1469–77, 2004.
33. Wolf, P. D. and Nicolelis, M. A. L. National Institutes of Health, http://www.ninds.nih.gov/funding/research/npp/niw06_poster_abstracts.pdf.
34. Puers, R. and Vandevoorde, G. Recent progress on transcutaneous energy transfer for total artificial heart systems. *Artif. Organs* 25(5), 400–405, 2001.
35. Pennes, H. H. Analysis of tissue and arterial blood temperatures in the resting forearm. *J. Appl. Physiol.* 1(2), 93–122, 1948.
36. Lazzi, G. Thermal effects of bioimplants. *IEEE Eng. Med. Biol. Mag.* 24(5), 75–81, 2005.
37. Kiyatkin, E. A. Brain hyperthermia during physiological and pathological conditions: causes, mechanisms, and functional implications. *Curr. Neurovasc. Res.* 1, 77–90, 2004.
38. IEEE Standard for Safety Levels With Respect to Human Exposure to Radio Frequency Electromagnetic Fields, 3 kHz to 300 GHz. In IEEE Standard C95.11999.
39. Matsuki, H., Matsuzaki, T., and Satoh, T. Simulations of temperature rise on transcutaneous energy transmission by non-contact energy transmitting coils. *IEEE Trans. Magnetics* 29(6), 3334–36, 1993.
40. Geselowitz, D. B., Hoang, Q. T., and Gaumond, R. P. The effects of metals on a transcutaneous energy transmission system. *IEEE Trans. Biomed. Eng.* 39(9), 928–34, 1992.
41. Tang, Q., Tummala, N., Gupta, S. K. S., and Schwiebert, L. Communication scheduling to minimize thermal effects of implanted biosensor networks in homogeneous tissue. *IEEE Trans. Biomed. Eng.* 52(7), 1285–94, 2005.
42. Gabriel, S., Lau, R. W., and Gabriel, C. The dielectric properties of biological tissues. II. Measurements in the frequency range 10 Hz to 20 GHz. *Phys. Med. Biol.* 41 (11), 2251–69, 1996.
43. Ibrahim, T. S., Abraham, D., and Rennaker, R. Electromagnetic power absorption and temperature changes due to brain machine interface operation. *Ann. Biomed. Eng.*, 2007.
44. Christ, A., Samaras, T., Klingenböck, A., and Kuster, N. Characterization of the electromagnetic near-field absorption in layered biological tissue in the frequency range from 30 MHz to 6000 MHz. *Phys. Med. Biol.* 51, 4951–65, 2006.
45. Kuster, N. and Balzano, Q. Energy absorption mechanism by biological bodies in the near field of dipole antennas above 300 MHz. *IEEE Trans. Veh. Technol.* 41(1), 17–23, 1992.
46. Samaras, T., Christ, A., and Kuster, N. Worst case temperature rise in a one-dimensional tissue model exposed to radiofrequency radiation. *IEEE Trans. Biomed. Eng.* 54(3), 492–496, 2007.
47. Kim, J. and Rahmat-Samii, Y. Implanted antennas inside a human body: simulations, designs, and characterizations. *IEEE Trans. Microwave Theory Techniques* 52(8), 1934–43, 2004.
48. Lazzi, G., DeMarco, S. C., Wentai, L., Weiland, J. D., and Humayun, M. S. Computed SAR and thermal elevation in a 0.25-mm 2-D model of the human eye and head in response to an implanted retinal stimulator. Part II. Results. *IEEE Trans. Antennas Propagation* 51(9), 2286–95, 2003.

49. Jain, M. K. and Wolf, P. D. Temperature-controlled and constant-power radio-frequency ablation: what affects lesion growth? *IEEE Trans. Biomed. Eng.* 46(12), 1405–12, 1999.

50. Seese, T. M., Harasaki, H., Saidel, G. M., and Davies, C. R. Characterization of tissue morphology, angiogenesis, and temperature in the adaptive response of muscle tissue to chronic heating. *Lab. Invest.* 78(12), 1553–62, 1998.

51. Davies, C. R., Fukumura, F., Fukamachi, K., Muramoto, K., Himley, S. C., Massiello, A., Chen, J. F., and Harasaki, H. Adaptation of tissue to a chronic heat load. *ASAIO J.* 40(3), M514–17, 1994.

52. Okazaki, Y., Davies, C. R., Matsuyoshi, T., Fukamachi, K., Wika, K. E., and Harasaki, H. Heat from an implanted power source is mainly dissipated by blood perfusion. *ASAIO J.* 43(5), M585–88, 1997.

53. Kiyatkin, E. A., Brown, P. L., and Wise, R. A. Brain temperature fluctuation: a reflection of functional neural activation. *Eur. J. Neurosci.* 16(1), 164–68, 2002.

54. Goldstein, L. S., Dewhirst, M. W., Repacholi, M., and Kheifets, L. Summary, conclusions and recommendations: adverse temperature levels in the human body. *Int. J. Hyperthermia* 19(3), 373–84, 2003.

55. Haveman, J., Sminia, P., Wondergem, J., van der Zee, J., and Hulshof, M. Effects of hyperthermia on the central nervous system: what was learnt from animal studies? *Int. J. Hyperthermia* 21(5), 473–87, 2005.

56. Dewey, W. C. Arrhenius relationships from the molecule and cell to the clinic. *Int. J. Hyperthermia* 10(4), 457–83, 1994.

57. Dewhirst, M. W., Viglianti, B. L., Lora-Michiels, M., Hanson, M., and Hoopes, P. J. Basic principles of thermal dosimetry and thermal thresholds for tissue damage from hyperthermia. *Int. J. Hyperthermia* 19(3), 267–94, 2003.

58. Xie, T., McCann, U. D., Kim, S., Yuan, J., and Ricaurte, G. A. Effect of temperature on dopamine transporter function and intracellular accumulation of methamphetamine: implications for methamphetamine-induced dopaminergic neurotoxicity. *J. Neurosci.* 20(20), 7838, 2000.

59. Gao, B. O., Franken, P., Tobler, I., and Borbely, A. A., Effect of elevated ambient temperature on sleep, EEG spectra, and brain temperature in the rat. *Am. J. Physiol. Regul. Integr. Comp. Physiol.* 268(6), 1365–73, 1995.

60. Carlisle, H. J. and Ingram, D. L. The effects of heating and cooling the spinal cord and hypothalamus on thermoregulatory behaviour in the pig. *J. Physiol.* 231(2), 353–64, 1973.

61. Magoun, H. W., Harrison, F., Brobeck, J. R., and Ranson, S. W. Activation of heat loss mechanisms by local heating of the brain. *J. Neurophysiol.* 1(2), 101–14, 1938.

62. Lindquist, S. and Craig, E. A. The heat-shock proteins. *Annu. Rev. Genet.* 22(1), 631–77, 1988.

63. Arora, H., Klemmer, N., Morizio, J. C., and Wolf, P. D. Enhanced phase noise modeling of fractional-N frequency synthesizers. *IEEE Trans. Circuits Syst.* 52(2), 379–95, 2005.

Part III

4 *In Vitro* Models for Neuroelectrodes: *A Paradigm for Studying Tissue–Materials Interactions in the Brain*

Vadim Polikov, Michelle Block, Cen Zhang, W. Monty Reichert, and J. S. Hong

CONTENTS

4.1 THE IMPORTANCE OF *IN VITRO* MODELS

Currently, the detailed mechanisms responsible for the electrode-tissue interactions leading to the formation of the glial scar around chronically implanted neuron-recording electrodes are poorly understood. As a consequence, the long-term use of electrodes for human prosthetics largely remains an untapped therapeutic resource. In this chapter, we will define currently available *in vitro* models that have the potential to uncover how the glial scar impedes brain electrode function and that provide a test bed to screen potential therapeutic compounds. Further, we will discuss the strengths and weaknesses of each of the *in vitro* models and explain how this line of inquiry could advance the field.

Multiple cell types interact to result in normal brain physiology and the formation of the glial scar. Through *in vitro* ("in glass") studies, as opposed to *in vivo* ("in life") experiments, we may begin to directly address the effects of electrodes and foreign objects at a cellular and molecular level, clarifying the respective contribution of individual cell types and basic mechanisms. Specifically, cell culture studies offer the potential for detailed inquiry into how electrodes initiate the glial scar, what drives glial scar formation, and the glial scar mechanisms responsible for electrode signal degradation. Ideally, this process will reveal novel therapeutic strategies capable of extending the viability and utility of brain electrodes.

4.1.1 RELATIVE ADVANTAGES OF *IN VITRO* MODELS

The primary advantage of *in vitro* studies is that they offer a controlled environment to test specific cellular and molecular hypotheses (Table 4.1). Tissue culture controls for environmental influences (e.g., temperature) and physiologic conditions of the animal (e.g., hormones, illness, nutrition), result in reduced experimental variation and assist in the ease of experimental replication [1]. By using controlled combinations of relevant and specific cell types, it is possible to obtain valuable insight into how these critical cells interact. For example, using reconstituted cultures, it has been shown that lipopolysaccharide (LPS) is toxic to dopaminergic neurons only in the presence of microglia, the resident innate immune cell in the brain. While mixed cultures comprised of neurons, astrocytes, and microglia resulted in neuron damage, those cultures containing only neurons showed no LPS effect. However, addition of microglia back into neuron-enriched cultures reinstated the LPS-induced neurotoxicity, while addition of astrocytes did not result in any LPS-induced effect [2]. Thus, using cell culture, it becomes possible to identify critical cell types responsible for glial scar formation and reveal essential cell–cell interactions. In addition to extracellular interactions, culture allows inquiry into the biochemical processes responsible for electrode failure.

Another great advantage of tissue culture is the ease of experimental procedures, experimental designs, and shortened experimental timescales. Although animal dissections are required to acquire primary tissue, the time, expense, and animal treatment stress associated with *in vivo* work on electrode biocompatibility is eliminated by *in vitro* models. In addition, treatment (agonists, antagonists, or potential therapeutic compounds) is applied in lower volumes (milliliters), drastically reducing the cost of compounds when compared to whole animal exposures.

TABLE 4.1
General Benefits and Disadvantages of *In Vitro* Models

Benefits of Tissue Culture	Examples
Control of Environment	Temperature, pH, hormone and nutrient concentrations.
Control of Cellular Constitution	Defined cellular identity and controlled cellular interactions.
Less Time	Time measured in days to weeks, rather than months.
Scale: Less Reagent Required	Distribution of compounds in milliliter volumes compared to systemic distribution.
Less Expensive Than *In Vivo* Experiments	After initial equipment acquisition, the cost of consumables and media are significantly less expensive than animal husbandry.
Replicates and Variability	Traditionally easy to replicate and less variability than *in vivo* studies.
Reduction of Animal Use	Treatment is administered to cell lines or primary cultures/tissue rather than in the intact animal. Animals stress due to treatment is reduced.

Disadvantages of Tissue Culture	Examples
Lack of Systemic Input from the Periphery	Brain pathology receives input from the peripheral system.
Sterile Technique/Expertise	The process of cell culture requires knowledge about sterile technique and the cell types of interest.
Potential for Dedifferentiation and Selection	The cell types may not be identical to cells in the intact system.
Three Dimensional Structure Lost	The effects of tissue and organ structure are not present.

*Adapted from [1].

Together, these characteristics add to the convenience of *in vitro* experiments and work to attenuate the traditionally excessive expense associated with *in vivo* studies of biocompatibility.

4.1.2 *In Vitro* Model Limitations

All models are wrought with limitations that must be considered in overall experimental design (Table 4.1). The most widely recognized limitation of tissue culture is based on the fact that the cells grown *in vitro* are not the exact dissociated replicates of their *in vivo* counterparts. For example, dissociation may (or may not) change cell identity or function. In primary culture, often embryonic cells are used and younger less differentiated cells may have different phenotypes than mature *in vivo* counterparts. Furthermore, there is a known genetic and phenotypic instability in many cell lines and primary cultures. Finally, the artificial and controlled environment is not the same as what the cells experience *in vivo* and could have an effect, depending upon the measure being assessed. Whether tissue culture experiments will accurately mimic *in vivo* behavior will vary based on the cell types tested, the treatment being investigated, and the endpoints measured. However, the potential experimental downfalls can be identified and consequently accounted for with the proper use of both positive and negative controls to discern that the tissue culture model is functioning consistently with *in vivo* physiology. In addition, critical components of novel *in vitro* findings must be confirmed *in vivo*. Thus, while *in vitro* models are a powerful tool of mechanistic diagnosis and a convenient, cost effective method of screening, they cannot completely replace animal and human studies.

4.1.3   THE VALUE OF *IN VITRO* ANALYSIS TO EXPLAIN *IN VIVO* OBSERVATIONS: A CASE STUDY

While *in vivo* studies are the gold standard in biomaterial–tissue biocompatibility studies, *in vitro* studies have greatly contributed to our understanding of the mechanisms of numerous human diseases when a phenomenon observed *in vivo* is then mechanistically explored at the cellular and molecular level *in vitro*. As a case study, through the use of *in vitro* central nervous system (CNS) models, our lab and others were able to drastically redefine the role of microglia in the development and progression of neurodegenerative diseases. In this instance, microglial activation in neurodegenerative disease was initially perceived as only a compensatory and passive response with no real deleterious consequences. *In vivo*, it was evident that microglia were activated in neurodegenerative disease, because microglia were detected in postmortem analyses of human patients, such as in Alzheimer's disease [3] and Parkinson's disease [4]. More recent *in vivo* studies made it possible to non-invasively monitor microglial activation in neurodegenerative diseases in patients [5] and animal models of Parkinson's disease [6]. However, the *in vivo* experiments and studies of human patients were unable to answer why these microglia were activated in these neurodegenerative diseases and neuropathologies, and it was impossible to determine if the microglia cell type was culpable in the neuronal damage. Thus, to address these mechanistic questions, the focus had to center on the cellular level, and investigation had to shift to *in vitro*.

The research effort of the Hong laboratory focuses on the mechanism of the pathogenesis of Parkinson's disease, which is a neurodegenerative disease characterized by the selective and progressive loss of dopaminergic neurons in the substantia nigra brain region. Using *in vitro* models, we were able to reproduce the *in vivo* hallmarks of Parkinson's disease etiology and gain further insight into the mechanisms behind disease progression by showing (a) prominent inflammation resulting in neuronal death, and (b) delayed, progressive, and selective loss of dopaminergic neurons, similar to what occurs in Parkinson's disease. We used a variety of culture systems to explore these phenomena. We used cell lines in biochemical assays when large amounts of protein or genetic material are needed to run Western blots or real-time polymerase chain reaction (PCR) assays to look at intracellular signaling pathways. However, cell lines are only used after the behavior is observed in primary cultures since results seen with cell lines but not seen *in vivo* or in primary cultures are not trustworthy.

Since progressive neurodegeneration is the result of complex interaction between microglia, astrocytes, and neurons, most of our experiments begin with the neuron–glia culture, an embryonic culture that contains a physiologically relevant mix of these three cell types. To probe the mechanisms further, we developed reconstituted primary cultures (neuron enriched, microglia enriched, astroglia enriched, microglia depleted) where specific cell types are depleted from neuron–glia cultures and added back in an effort to identify the contribution of individual cell types and how these cell types interact with each other. These reconstituted cultures gave us more control over the experimental variables by removing specific cell type interactions (i.e., neuron–microglia interactions are gone in microglia-depleted cultures), although they resembled the *in vivo* condition less than the complex neuron–glia culture. Thus, only through cell culture were we able to demonstrate that microglia were capable of actively inducing neuron damage when activated [2]. Using this same approach, we were also able to show that microglia play a role in 1-methyl 1,4-phenyl 1,2,3,6 tetrahydropyridine (MPTP)-induced neuronal damage [7]. Thus, through *in vitro* cell culture, we were able to show that what was previously assumed to be a casual response (microglial activation) when observed *in vivo* was in fact causing neuron damage. This line of inquiry was important because this *in vitro* model has helped us understand more about the mechanisms of microglia-mediated neurotoxicity in response to endogenous and environmental toxins, and it also provides valuable insight into novel therapy through an antiinflammatory approach.

Clearly, *in vitro* models provide important information that, when combined with *in vivo* validation and a respect for the limitations of *in vitro* analysis, can help the neuroprosthetic community understand and design around biocompatibility failures of chronic brain implants.

4.2 TYPES OF MODELS

A variety of *in vitro* models involving living cells have been employed to look at material–tissue interactions that might occur in the brain. These models range widely in complexity from very simple single cell lines to complex organotypic brain slices. As the complexity of the model increases, the relevance of the data to the actual *in vivo* conditions also increases, but the tradeoff is a loss of predictability, simplicity, and control. One possible decision tree with this tradeoff in mind is presented as Figure 4.1. The decision of which *in vitro* model to employ should be made with the aim of the experiment and the limitations of each system in mind. For example, cytotoxicity studies looking at the effect of monomer leaching from a polymeric implant would be better served by a fibroblast cell line rather than a primary cell culture composed of many interacting cell types. The fibroblast cell line is easier to culture, cheaper, and has an established experimental protocol that has been used in hundreds of cytotoxicity studies, resulting in established controls. However, if neurons are known to be more sensitive to a particular pharmacological insult (such as the leaching monomer) than fibroblasts, then primary neuronal

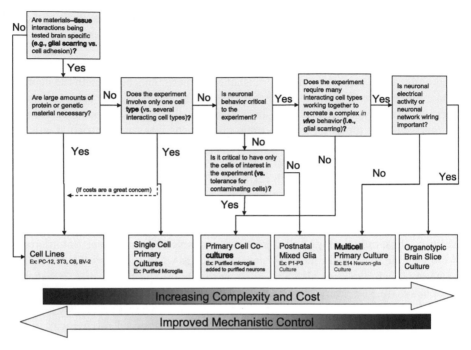

FIGURE 4.1 One of many possible decision trees that can be used to determine the best *in vitro* model to use when planning an experiment.

cultures may be more appropriate. However, if a new protocol is chosen over an established design, the new design requires greater characterization and validation of controls.

In the rest of this chapter, we will review the different *in vitro* systems employed by investigators to study neuroelectrode–tissue interactions and attempt to offer insights into future models currently being developed within our own lab. We begin the review with simple cell line models, proceed to primary single-cell models and then more complex primary culture systems with interacting cell types, and finally end with organotypic tissue slices, which represent an intermediate model between dissociated cell culture and *in vivo*.

4.2.1 CELL LINES

The simplest *in vitro* models involve cultures of immortalized cell lines. There are several advantages to the use of cell lines: they are easy to culture, easy to grow and maintain indefinitely, inexpensive relative to primary cultures, and generate highly reproducible results. For example, it is much easier to obtain large amounts of protein or mRNA for analysis, particularly when compared to primary cultures. Furthermore, cell lines are derived from a single cell type, so there is no danger of contamination by other confounding cell types. However, the limitations of cell lines are severe and must be well appreciated to accurately interpret experimental results. Since cell lines are derived from tumors or other genetically unstable cells, the cell line phenotype will only approximate the normal, genetically stable cell population. For example, the emphasis of the cell line's metabolism may have shifted to growth and proliferation rather than the resting, fully differentiated function of the cell type that the cell line is approximating [8]. The genetic instability of cell lines also contributes to a loss in reproducibility and relevance over multiple passages. In addition to noting the source of your cells and the passage number, it is usually best practice to keep the passage number variance to a minimum in experiments. When these limitations are combined with the more general limitations of two-dimensional dissociated cell cultures, it is often difficult to place much faith in those results observed in cell line studies but not yet confirmed *in vivo* or in primary culture. However, as long as the limitations of immortalized cells are kept in mind, cell lines can be a powerful tool in the *in vitro* arsenal.

4.2.1.1 Fibroblast Cell Lines

The simplest biocompatibility experiments that can be conducted are cytotoxicity and nonspecific cell adhesion assays. Stephen Massia's group regularly uses the 3T3 fibroblast cell line to evaluate the cytotoxicity of new neuroelectrode materials and coatings [9–11]. This cell line is among a set of well-characterized lines mandated for use by the FDA as part of the cytotoxicity testing protocols for approving new medical devices. 3T3 fibroblasts arc grown in a monolayer and exposed to materials such as polyimide- and silicon-based neuroelectrodes [9], diamond-like carbon coatings [10], and bioactive dextran/peptide coatings on a model substrate [11] (Figure 4.2). A standard live–dead assay is conducted after 24 hours to see if the material tested caused any cytotoxicity. For adhesion studies, cells are seeded on top of materials, and the number of cells adhering to the material after 24 hours is counted and compared to a base material or tissue culture polystyrene. Fibroblast cell lines are often used because they are easy to grow and maintain. However, since the meninges contain fibroblast-like cells, sometimes fibroblast cell properties are desired when comparing adhesion to brain cell types such as neurons and astrocytes. For example, 3T3 fibroblast adhesion was compared to astrocyte and neuronal cell line adhesion on various substrates. The degree of fibroblast cell line adhesion was found to be significantly different from that of a neuronal cell line on RGD adhesion peptide modified surfaces but similar to that of a glial cell line [11]. In contrast, a different group using a different fibroblast cell line derived from rat skin (CRL-1213; American Type Culture Collection [ATCC]) did show a significant difference in adhesive properties between the fibroblast and glial cell lines on RGDS-modified surfaces [12], underscoring the variability in responses between different cell lines.

4.2.1.2 Neuronal Cell Lines

One line of neuroelectrode research has focused on finding materials that attract neuronal processes or stimulate neurite growth. For these experiments, neuronal cell lines have often been used and their responses to various materials compared to other cell types found in the brain. The most widely used neuronal cell line in the neurobiology community is the PC-12 cell line, which exhibits many neuronal

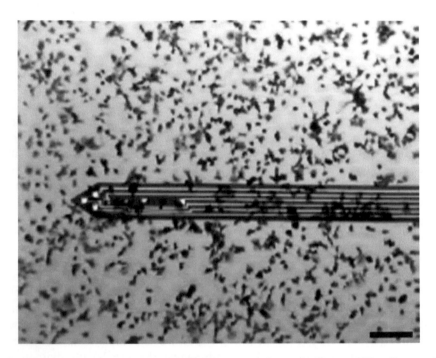

FIGURE 4.2 Photomicrograph depicting the morphology of adherent 3T3 cells on a PI electrode shank and surrounding wafer surface (scale bar = 100 μm). (Adapted from Lee, K. K., He, J. P., Singh, A., Massia, S., Ehteshami, G., Kim, B., and Raupp, G., *J. Micromech. Microeng.*, 14, 32–37, 2004. With permission.) **(See color insert following page 110.)**

behaviors, especially reversible differentiation and neurite growth in response to NGF. A typical experiment with this cell line involves coating the substrate of interest with laminin or another molecule to promote adhesion, seeding the cells on the substrate, waiting for cell growth and differentiation (typically 1 to 3 days), and then fixing the cells and using immunostaining to image cell bodies and neurites. To measure adhesion, the number of cells is counted in random microscope fields to obtain a cell number or cell coverage ratio. A more sensitive measure to look at how the electrode material (cell culture substrate) affects neuronal growth involves counting the neurites and even measuring average neurite length as the immortalized neuronal cells differentiate. In two such experiments, PC-12 cell neurite growth was found to be more robust on textured silicon neuroelectrodes than on nonporous silicon by Moxon et al. [13], and PC-12 cells were found to adhere to IKVAV peptide modified polyimide and silicon oxide surfaces to a much greater extent than fibroblast or glial cell lines [11]. Martin's group is attempting to integrate a conducting polymer grown from the neuroelectrode recording sites with neurons around the electrode to try to establish a more reliable and robust signal. They found that after seeding the cells on top of several different substrates, the human neuroblastoma SH-SY5Y cell line preferentially grew on top of the polypyrrole/CDPGY-IGSR polymer/peptide blend "grown" from the recording sites of a silicon electrode [14]. Over a dozen well-characterized neuronal cell lines are available as a result of the decades of cell line experiments in the neurosciences [8].

4.2.1.3 Astrocyte Cell Lines

Because of the widespread understanding that the biocompatibility of neuroelectrodes is a function of the glial response to the implanted material, the majority of cell line–based experiments have been conducted using cell lines that approximate microglial or astroglial function. None of the astrocyte cell lines are perfect in recreating primary astrocyte behavior, but they are seen as more relevant than fibroblast cell lines in studying brain biocompatibility. As with neuronal cell lines, the most common experiment measures cell adhesion and growth of cells on a specific substrate. Cells are seeded in tissue culture plates containing the substrate of interest and allowed to adhere and grow for 1 to 3 days. The cells are then fixed, rinsed, and stained to assess the number of cells adhering to the substrate (as compared to other substrates or tissue culture plastic control). Cell spreading can also be measured to identify materials that promote a specific type of behavior (whether one desires spreading or no spreading depends on the objectives of the experiment).

One of the most common astroglial cell lines is the C6 line derived from a rat N-nitrosomethylurea-induced glioma. The cell line expresses one stereotypical astroglial marker, S100, but not the most specific astrocyte marker, glial fibrillary acid protein (GFAP). C6 cells were used to show preferential glial cell adhesion to polypyrrole/silk-like having fibronectin fragments (SLPF) polymer/peptide blend on a silicon electrode [14] (Figure 4.3). The DITNC1 astrocyte cell line (ATCC CRL-2005) expresses GFAP and appears to be similar in phenotype to type 1 astrocytes in culture (type 1 astrocytes are more physiologically relevant, while type 2 astrocytes are thought to be an *in vitro* phenomenon). This cell line showed lower adhesion to textured neuroelectrodes than to smooth silicon electrodes [13], which together with the neurite outgrowth data presented above suggests texturing as a possible strategy to improve biocompatibility. Massia's group uses another Fischer rat glioma line, F98 (ATCC #CRL-1690) in experiments testing astrocyte adhesion to different surfaces [10,11].

Several groups working on neuroelectrode biocompatibility at Cornell University, the Wadsworth Center, and Rensselaer Polytechnic Institute have used the LRM55 astroglial cell line to test new materials and modifications. A transformed rat glioma cell line originally developed by Martin and Shain [15], this cell type has been shown to exhibit several important astrocyte metabolic features. Investigators have used this cell line to show adhesion and spreading on micropatterned hydrophilic hexagonal trimethoxysilyl-propyl-diethylene triamene (DET)A surfaces of varying sizes [16], an LRM55 preference for microfabricated silicon pillars and wells [17], the cell line preference for microcontact printed surfaces over traditional photolithographic patterned surfaces [18], and a preference for smooth chemically etched regions over roughened "silicon grass" regions [19] (Figure 4.4). Contradictory behavior from primary astrocytes, which favor roughened silicon grass regions, underscores the limited ability to make definitive conclusions from cell line experiments [17].

4.2.1.4 Microglia Cell Lines

Microglia-based cell lines are commonly used to study neuroinflammatory processes, but only recently have biocompatibility studies recognized the dominant

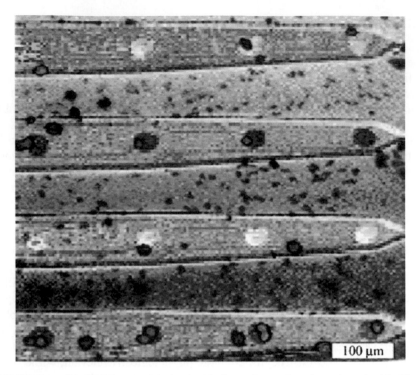

FIGURE 4.3 PPy/SLPF-coated 4-shank, 16-channel neural probe cultured with C6 cells. Cells were stained using Hoechst 33342, and the blue spots correspond to individual cells. (Adapted from Cui, X. Y., Lee, V. A., Raphael, Y., Wiler, J. A., Hetke, J. F., Anderson, D. J., and Martin, D. C., *J. Biomed. Mater. Res.* 56, 261–272, 2001. With permission.) **(See color insert.)**

role of microglial cells. Often cell lines are employed to better understand processes already known to occur *in vivo* or in primary cells. For example, Tzeng and Huang used the immortalized mouse microglial BV-2 cell line to probe the effect of neurotrophin-3 (NT-3) on the cytokines and inflammatory molecules released by microglia after an LPS immune challenge [20], a process suggested by *in vivo* observations. Kremlev et al. used the BV-2 cell line as well as the HAPI microglial-like rat cell line to assay for the release of chemokines and the expression of chemokine receptors after an immune challenge after *in vivo* data suggested the involvement of chemokines in inflammation-mediated brain injury [21]. Both the BV-2 and HAPI cell lines are commonly used in the Hong lab to further explore behaviors observed *in vivo* or in primary cultures [22]. Once the cell lines show the same behavior as we earlier observed *in vivo* (i.e., release of a certain cytokine after immune challenge), we can then use the cell lines to examine the details of that behavior (i.e., the signal transduction pathway of cytokine expression). Another example where microglial cell lines were advantageous over primary culture is in the case of studying LPS internalization in microglia. All microglia, including cell lines, phagocytose LPS, and it was simpler to use the cell line, because they were less sensitive than primary cultures to the lower-density seeding that was necessary to get single cell images for confocal microscopy (Figure 4.5) [23].

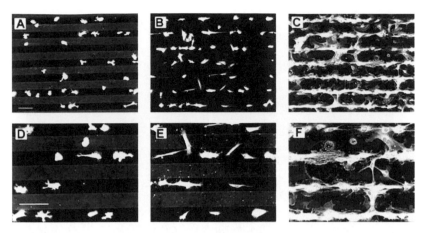

FIGURE 4.4 Time course of LRM55 astroglial cell attachment to surfaces patterned by microcontact printing. Cells were plated and fixed after 2 hours (A and D), 6 hours (B and E), and 24 hours (C and F). Bar, 100 mm for both low magnification (10× objective, A through C) and high magnification (20× objective, D through F) images. Dark regions are a permissive DETA (a hydrophobic organosilane self assembled monolayer), and the light stripes are layers of inhibitory OTS (a hydrophobic organosilane). (Adapted from St. John, P. M., Kam, L., Turner, S. W., Craighead, H. G., Issacson, M., Turner, J. N., and Shain, W., *J. Neurosci. Methods* 75, 171–177, 1997. With permission.)

4.2.2 PRIMARY CELLS

The step up in complexity from cell lines to primary cultures is significant. Primary cells are genetically stable and are often the actual cells taking part in the *in vivo* process, so the results from primary culture experiments are much more trustworthy. However, in exchange for the increase in relevance comes significant new challenges, including additional costs of animal purchase and husbandry, isolation difficulties, and increased culture variability. One also loses the ability to easily culture and proliferate cells through many generations. As primary cells proliferate and are passaged, the phenotype may change to a point where the cells no longer behave in the same way that the initially isolated cells behaved. For this reason, only the first or the first few passages of primary cells can be used, and the number of valid passages depends on the cell type and behavior being investigated.

Even with greater relevance to the *in vivo* situation, primary cell culture is still not a perfect model of the *in vivo* environment, as the isolation procedure often injures or activates the cell, resulting in an altered phenotype, and the two-dimensional culture system without vasculature, extracellular matrix (ECM), and other supporting cells further alters cellular behavior. Furthermore, as with any phenotypically diverse biological system, the results obtained with primary cultures will show a higher degree of variability than the results from genetically identical cell line cultures.

Primary cultures can be subdivided further into increasing levels of complexity. The least complex cultures are single cell type primary cultures. These are the primary cell equivalent of cell lines, where the researcher is interested in the response

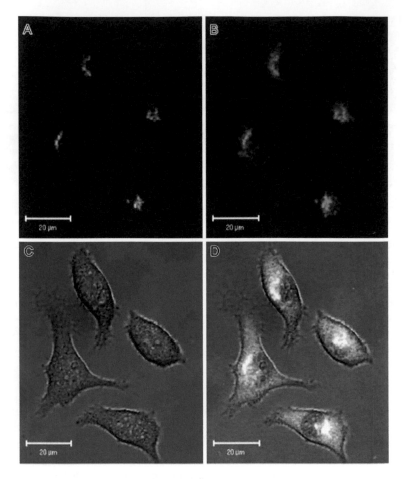

FIGURE 4.5 Internalized LPS is localized in the golgi. HAPI microglial cells were exposed to 10 µg/ml labeled LPS and 1 µM NBD-ceramide for 1 hour at 37°C. Perinuclear colocalization of NBD-ceramide (A) and Alexa568-LPS (B) in HAPI microglia. (C) The transmitted light image and (D) the overlay of transmitted light and fluorescence images. The scale bars indicate 20 µm. **(See color insert.)**

of a single cell type to the experimental conditions. By combining two independently isolated sets of primary cells (i.e., primary astrocytes and primary neurons), a more complex coculture can be used, thus allowing analysis of cell–cell interactions. Finally, if two or more cell types are isolated together from the same tissue in the same procedure (i.e., a culture of astrocytes and microglia from postnatal rat cortex), this is an additional step toward the more complex conditions of the *in vivo* environment.

4.2.2.1 Single Cell Type Primary Cultures: Primary Neurons

Because of their unique electrical properties relative to other cells, and their varied phenotypes within the CNS and peripheral nervous system (PNS), neurons are not well represented by cell lines in culture. While a few neuronal cell lines, such as the PC-12 line, exhibit limited neuronal behavior, the vast majority of neurobiological studies in the past several decades have used primary neuronal cultures. Neurons are highly varied in form and in behavior depending on their age, location, and function in the brain, but commonly share negligible to no proliferative potential, making cell culture quite difficult. Still, several isolation protocols have been established that consistently produce highly pure, electrophysiologically active neuronal cultures.

Because of the limited proliferative potential of adult neurons and their relative sensitivity to injury, neurons are usually isolated from embryonic or early postnatal animals, while they are still differentiating from the hardier, dividing neuronal precursor cells. Cells are mechanically or enzymatically dissociated, plated on poly-D-Lysine or laminin-coated polystyrene, and maintained in specially defined, serum-free media. Dobbertin et al. used a standard cortical neuronal culture from embryonic day 17 (E17) Sprague–Dawley rats to study the growth of neurites on RPTPβ/phosphacan, a chondroitin sulphate proteoglycan upregulated after injury to the CNS [24]. After removal of the striatum and hippocampus, the cerebral cortices were freed of meninges and cut into small pieces before enzymatic treatment with trypsin. By seeding at low density (10^4 cells/cm^2 and using specially defined growth media (1:1 mixture of DMEM and Ham's F12 with 1% N2 supplement), the investigators were able to get a cell culture that was 98% pure neurons. Ravenscroft et al. used a similar procedure to isolate E18-19 hippocampal neurons in their studies and patterned them on silane modified surfaces [25]. They were then able to engineer circuits of aligned neurons on glass coated with a patterned silane film. In this case, cells were dissociated with papain and layered over a step gradient to remove cellular debris before resuspension in serum-free neurobasal medium supplemented with B27, glutamine, and glutamate.

Postnatal-derived neuronal cultures have a lower neuronal yield since most neurons perish in the isolation process and require more stringent precautions for removing nonneuronal cells, such as a nylon mesh to remove meninges and treatment with mitotic inhibitors to prevent the culture from being overgrown by proliferating glia [26]. Other neuronal cultures include embryonic spinal motorneuron cultures, embryonic or perinatal rodent sympathetic ganglia cultures, and embryonic dorsal root ganglion cultures [8].

If neurons are cultured alone without glia, they will become spontaneously electrophysiologically active within a week of the dissection and can remain viable and active in culture with the proper maintenance for many weeks to months. Potter's group and others have cultured neurons on microelectrode array (MEA) recording systems for over a year by maintaining the osmolality of the culture within a narrow range [27]. However, while much of the neuroscience community uses primary neuronal cultures, neuroelectrode engineers have not used them for biocompatibility studies because there is little need for accurate electrophysiological behavior when straightforward cytotoxicity and cell attachment assays are being performed. As greater emphasis is placed on more complex *in vitro* models (i.e., simultaneous observation of signal degradation and glial scar formation *in vitro*), and electrically active electrodes are used in culture, primary neurons will need to replace neuronal cell lines in future biocompatibility studies.

4.2.2.2 Single Cell Type Primary Cultures: Primary Glia

Primary glial cultures are easier to generate than primary neuronal cultures because glia have much higher proliferative abilities and are more likely to be part of the approximately 1% of cells that survive the isolation procedure [8]. Furthermore, since the diversity of astrocytes, microglia, and oligodendrocytes is not well understood, unlike the much greater appreciation for neuronal phenotype differences (i.e., dorsal root ganglion versus motorneuron isolation), researchers typically use heterogeneous cultures of glia. A procedure to isolate glia initially established by McCarthy and de Vellis in 1980 has been modified by various groups but has roughly remained intact for over 25 years [28]. In this procedure, perinatal (postnatal day 1 to 4) rat or mouse cerebral brain cells are plated at high density ($2 \times 10^5/cm^2$) in serum-supplemented medium after removal of the meninges and mechanical or enzymatic dissociation. At this point, glia are still dividing and, with the help of serum and endogenously produced cytokines, will create a confluent layer of astrocytes, glial precursor cells, and microglia within 6 days. The astrocytes in these cultures tend to form a confluent layer on top of the poly-D-lysine–coated polystyrene, with the loosely adherent microglia resting on top of the astrocyte layer. To isolate primary cultures of microglia, researchers take advantage of these differences in adherence and shake the cultures two weeks after plating and then again three weeks after plating. Although the culture can be shaken further, the microglia from additional shakes are no longer used in order to avoid experimental error derived from clonal expansion of particularly adherent microglia.

The media are collected after each shake and centrifuged, resulting in microglia that are around 95% pure and an astrocyte layer that is more than 98% pure. Shaking occurs on a standard rotary shaker at room temperature, although the times and speeds vary from lab to lab (our lab shakes for less than 3 hours at 180 RPM). The longer and faster the cells are shaken, the purer the astrocyte culture will be. However, shaking also activates the microglia, so longer shake times may result in more basal microglial activation. Astrocytes grown this way can survive for many weeks and can be subcultured several times, while microglia cannot survive for more than a few days without the supporting astrocyte layer. This method by McCarthy and de Vellis forms the basis for the primary glial cell isolation procedure, but each research group tends to make adjustments based on their own experience. Each group varies details such as the shake date, duration, speed, and frequency; the media; and the serum formulation.

4.2.2.3 Single Cell Type Primary Cultures: Primary Microglia

Our laboratory routinely maintains primary enriched-microglia cultures, which are prepared from the whole brains of 1-day-old Fisher 344 rat pups or mice according to a variation of the McCarthy and de Vellis procedure [29]. After removing meninges and blood vessels, the brain tissue (minus olfactory bulbs and cerebellum) is gently triturated and seeded (5×10^7, approximately 2.5 rat pup brains or 4 mouse pup brains) in 150-cm³ flasks. The culture medium consists of DMEM-F12 media supplemented with 10% heat-inactivated fetal bovine serum (FBS), 2 mM L-glutamine, 1 mM sodium pyruvate, 100 mM nonessential amino acids, 50 U/ml penicillin, and 50 lg/ml streptomycin. Cultures are maintained at 37°C in a humidified atmosphere of 5% CO_2 and 95% air. One week after seeding, the media is replaced. Two weeks after seeding, when the cells reach a confluent monolayer of glial cells, microglia are shaken off. Afterward, cells are resuspended in a treatment media (DMEM-F12 media supplemented with 2% FBS, 50 U/ml penicillin, and 50 ug/ml streptomycin) and replated at 1 × 10⁵ in a 96-well plate. Cells are treated 12 to 24 hours after seeding the enriched microglia. Media can be replaced in the original 150-cm³ flask of mixed glia, and a week later, an additional shake of microglia can be harvested.

Other than this standard culture preparation, there are other preparations that are less commonly used. Microglia that are 92 to 97% pure can be bulk isolated from older rats, as described Basu et al., who used a series of digestive incubations, nylon meshes, and centrifugation steps to isolate microglia from 8 to 12 week Sprague–Dawley rats [30]. Other cells known to participate in the glial scarring reaction, such as meningeal cells [26] and O2A precursor cells [24], can be isolated but have not yet been used widely in biocompatibility experiments.

4.2.2.4 Single Cell Type Primary Cultures: Primary Astrocytes

The astrocyte isolation procedure is the inverse of the microglia isolation procedure since microglia are shaken off and discarded while the astrocytes are maintained and subcultured. Dobbertin et al. used 2-day-old Sprague–Dawley rats, enzymatically triturated small pieces of tissue using 0.1% trypsin in HBSS for 20 minutes, and seeded poly-D-lysine-coated flasks with 10% fetal calf serum in DMEM [24]. The cultures were shaken between the 8th and 12th days, and the cells were subcultured and treated with 20 µM of the antimitotic cytosine-1-β-D arabinofuranosid (Ara-C) in serum-free media to remove any remaining proliferating glial precursors. Remaining microglia were removed by treatment with 10 mM L-leucine methyl ester (LME), which is a phagocyte toxin.

The more shakings that the astrocytes go through, the more microglia are removed, and a purer astrocyte culture is generated. After weekly shakings for 6 weeks, a 98% pure astrocyte culture remains. However, this method will produce a significant number of type 2 astrocytes, so our group prefers to acquire astrocytes using a primary cortical astrocyte method, as previously described [31] with a slight modifications [32]. Again, whole brains from 1-day-old rats are isolated, and the meninges are removed. However, for astrocytes, the cerebral cortices are dissected and subjected to enzymatic digestion for 15 min in DMEM/F-12 containing 2.0 mg/ml porcine trypsin and 0.005% DNase I. The tissue is then mechanically disaggregated using a 60-µm cell dissociation kit (Sigma-Aldrich, St. Louis, Missouri) to yield a mixed glial cell suspension. The cell suspension is centrifuged for 10 min at 300 g and resuspended in fresh culture medium: DMEM/F-12 supplemented with 10% FBS, 100 µM nonessential amino acids, 100 µM sodium pyruvate, 200 µM L-glutamine, 50 U/ml penicillin, and 50 µg/ml streptomycin. The cells are plated on

75-cm^2 polystyrene tissue culture flasks and maintained at 37°C, 5% CO_2, and 95% air until confluency for 2 weeks. Fresh medium is replenished every 3 to 4 days. Following confluency, the cells are placed on an orbital shaker at 150 revolutions/min for 6 hrs to remove contaminating cells (mostly microglia). The cells are harvested with 0.1% trypsin/EDTA in Hank's Balanced Salt Solution and plated in either T25 flasks or 100-mm Petri dishes at a density of 0.35 to 1 × 10^6 cells. Experimental studies are performed within 3 to 4 weeks of initial plating. Specifically, cells are treated when they are confluent, approximately 1 week after the last seeding. Upon treatment, cultures are switched to fresh medium containing 2% heat-inactivated FBS, 2 mM L-glutamine, 1 mM sodium pyruvate, 50 U/ml penicillin, and 50 µg/ml streptomycin.

4.2.2.5 Primary Cell Coculture

A step up in complexity from primary cultures of a single cell type is cocultures where two different cell types from two different sources are combined. The result is a culture that allows the study of cell–cell interactions between two cell types and is not limited by the different sensitivities of two different cell types to one isolation procedure. For example, sensitive neurons can be isolated by one procedure in an embryonic animal, while hardier astrocytes can be isolated by a different procedure in an adult animal. The two cell suspensions are seeded together, or one atop the other, and neuron–astrocyte interactions can be observed, or behaviors only observed in the presence of both cell types together can be tested. Although these cultures may be more relevant to the *in vivo* situation since cells are combined with other cell types that they normally interact with *in vivo*, relevance may be lost if different sources and isolation procedures result in behavior that is not physiologically relevant.

To test the effect of different materials in influencing cortical astrocyte ability to promote neuronal growth, Biran et al. seeded cerebellar granule neurons isolated from 7-day-old rats through a panning procedure atop a layer of astrocytes isolated from P-1 rat cerebral cortex that had been purified away from microglia and other cells by shaking (as described in the previous section) [33]. Since the P-1 dissection procedure does not yield neurons, yet neurons were important in assessing the growth-promoting properties of cortical astrocytes, the coculture was necessary to answer the question of biocompatibility posed by the researchers. They were able to use the culture to find no difference in astrocyte growth-promoting ability between different materials.

In one of the few attempts to look at astrogliosis *in vitro*, Guenard and colleagues used dorsal root ganglion (DRG) neurons prepared from E15 rats plated on top of astrocytes from E21 rat cortex or 2- to 3-month-old rat spinal cord to study the effect of mechanical axonal injury on astrocyte proliferation, GFAP expression, ECM deposition, and process orientation [34]. DRG neuron explants provide long-living neurons with large axons used to study axonal damage and regrowth after injury, whereas smaller neurons from the cortex (the astrocyte cell source) may have been difficult to isolate or analyze. Hirsch and Bahr seeded retinal explants from E16 rat embryos on top of adult optic nerve astrocytes or P-1 to P-3 cortical astrocyte layers to find that the retinal ganglion cell axons preferred to grow on astrocyte regions of the culture rather than the contaminating meningeal cell regions, putatively because of the different mix of inhibitory ECM expressed by the different basal layer cell types [35]. In another study, cortical neonatal astrocytes that were grown on a stretchable substrate, and thus aligned in specific orientations because of continuous mechanical strain in one direction, were used as the base layer to culture P-1 DRG neurons [36]. The study showed that aligned astrocytes also resulted in aligned ECM deposition and aligned neuronal process growth, thus potentially laying the groundwork for engineered glial substrates for spinal cord repair. Although astrocyte-neuron cocultures are most common, neurons can also be cocultured with other cell types, such as Schwann cells [34] and meningeal cells [35].

4.2.2.6 Multicell Primary Cultures: An *In Vitro* Model of Glial Scarring

The glial scarring response to implanted biomaterials involves at least three interacting cell types: neurons, astrocytes, and microglia. To keep the *in vitro* cell culture system as relevant to the *in vivo* environment as possible, all three cell types should optimally be isolated in the same isolation procedure from the same animal at the same time. Unlike experimenters who go to great lengths to isolate a specific cell type and remove "contaminating" cells, our approach has been to isolate and culture these interacting cell types together, thus more closely reproducing the *in vivo* environment. The benefit is that the complicated glial scarring response, which requires cell–cell interaction between at least three cell types, can be reproduced *in vitro* (Figure 4.6) [37]. We have been able to reproduce the glial scarring response to biomaterials placed in such culture, and the *in vitro* approach allows us to create a time course of events leading to glial scarring while potentially recording from the culture at the same time. We plan to further use the *in vitro* model to explore the mechanisms behind glial scar formation and further develop the model as a way to test new biocompatibility approaches that become available in the neuroprosthetics field.

The increased complexity also comes with several drawbacks as compared to simpler cell line or single cell type primary cell culture. First, the variability between culture preparations and experimenters is greater as more parameters can potentially be varied. The cell–cell interactions are more difficult to observe and analyze, as seven interactions are now possible between the three different cell types (A–A, B–B, C–C, A–B, A–C, B–C, A—B–C), whereas only one interaction is present in single cell type cultures (A–A), or three possible interactions are present in less complex cocultures cultures (A–A, B–B, A–B). Furthermore, since all the cell types have to be present in one isolation procedure, early embryonic (E14 to E17) cells are used to keep neurons alive, thus potentially reducing the culture's relevance to adult *in vivo* behavior.

Understanding the tradeoffs involved in using a more complex culture, we have decided that the benefits—namely the re-creation of the glial scar around electrode materials *in vitro*—outweigh the limitations in the system described above. The neuron–glia cell culture system contains all of the cell types relevant to glial scar formation. It is an embryonic day 14 midbrain culture that contains the physiologically relevant mix of approximately 40% neurons, 10% microglia, and 50% astrocytes

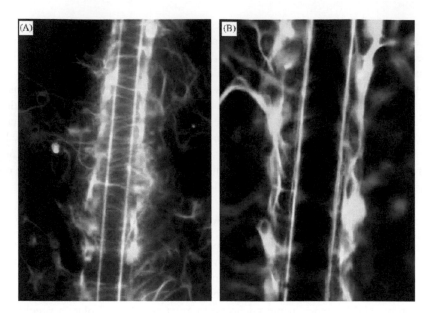

FIGURE 4.6   (A through B) Triple fluorescent labeling of a model electrode in neuron–glia culture with DAPI staining nuclei blue, GFAP staining green, and OX-42 staining microglia red shows the relative positions of different cells near the wire after 10 days in culture. Just as observed *in vivo*, there is a layer of microglia (red) adjacent to the microwire and astrocytes (green) outside of the microglial layer showing upregulated GFAP. The image in (B) shows the glial scarring at a higher magnification, clearly visualizing the prevalence of microglia around the microwire. For reference, the wire diameter is 50 μm in all images. (Adapted from Polikov, V. S., Block, M. L., Fellous, J. M., Hong, J. S., and Reichert, W. M., *Biomaterials* 27, 5368–5376, 2006. With permission.) **(See color insert.)**

[38]. To generate the culture, first, the midbrain is dissected out and the meninges removed. Then, the cells are mechanically triturated with various sized pipette tips and seeded at 5×10^5/well into poly-D-lysine-coated 24-well plates. Cells are maintained at 37°C in a humidified atmosphere of 5% CO_2 and 95% air, in minimal essential medium (MEM) containing 10% FBS, 10% horse serum (HS), 1 g/l glucose, 2 mM L-glutamine, 1 mM sodium pyruvate, 100 μM nonessential amino acids, 50 U/ml penicillin, and 50 μg/ml streptomycin. Seven-day-old cultures are used for treatment after a media change to MEM containing 2% FBS, 2% HS, 2 mM L-glutamine, 1 mM sodium pyruvate, 50 U/ml penicillin, and 50 μg/ml streptomycin.

Treatment involves several different interventions to induce the culture to reproduce the scarring or inflammatory behavior that leads to recording electrode failure *in vivo* [37]. One simple treatment option is to add 10 ng/ml of LPS into the culture, which results in microglial activation, inflammatory cytokine release, and neuronal bystander damage. To simulate physical damage and cell death characteristic of the mechanical injury sustained during device insertion, a scrape model is used (Figure 4.7A). In this model, an area of the confluent cell layer is scraped off with a cell scraper or a pipette tip, and cells are monitored as they repopulate the damaged region. A treatment to reproduce the foreign body response and glial scarring

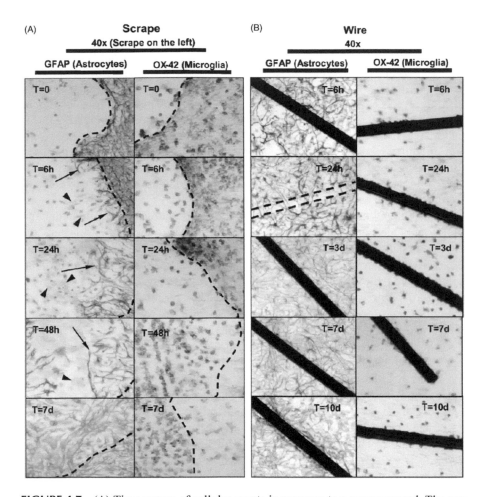

FIGURE 4.7 (A) Time course of cellular events in response to a scrape wound. The area scraped free of cell is on the left of the dotted line. Astrocytes are seen to send processes (arrows) into the wound beginning at 6 hours and continuing through 48 hours, and completely recolonize the wound by 7 days. GFAP-negative spindle-shaped precursor cells (arrowheads) that do not stain for microglial markers but stain for vimentin (not shown) migrate into and colonize the wound ahead of the GFAP-positive processes. Microglia migrate to and spread out within the wound by 24 hours, and their numbers increase over time, until by 7 days there are more microglia inside the wound than in the surrounding culture. (B) Time course of cellular events in response to the wire placement. Microglia attach to the wire as early as 6 hours and increase in number until a layer of microglia 1 to 2 cells thick is formed covering the length of the wire. This layer remains through 10 days in culture. Astrocytes show no response to the microwire until 10 days after treatment, a layer of activated astrocytes with upregulated GFAP forms around the microwire, mimicking the glial scarring seen *in vivo*. (Adapted from Polikov, V. S., Block, M. L., Fellous, J. M., Hong, J. S., and Reichert, W. M., *Biomaterials* 27, 5368–5376, 2006. With permission.) **(See color insert.)**

around an implant that occurs as a result of chronic electrode implantation involves the placement of short (2 to 5 mm) pieces of microwire into the culture and observe the cell reaction to the foreign body (Figure 4.7B). We use stainless steel wire because it is inexpensive, easy to sterilize, and is used in some microwire recording arrays, although we have used other materials with little observable difference in cellular response. Each culture is fixed at a certain time point after treatment and immunostained for different cell markers to differentiate neuronal (MAP-2, NeuN cell markers), microglial (IBA-1, OX-42), and astrocyte (GFAP) behavior. For the LPS treatment, fixation is usually done in the first 3 days since the maximal cytokine response is at 24 hours. The scrape is looked at with a time course ranging from several hours to 7 days (when the "wound" is filled), and the microwire treatment is extended as long as the culture is stable, typically 10 days.

This culture system has been used routinely to study the neuroinflammatory processes underlying progressive neurological diseases such as Parkinson's disease and Alzheimer's disease [2,29,38–47]. Removal of each of the different cell types destroys our ability to follow these processes, as each cell type contributes something to the overall disease process. We encounter the same logic with reproducing glial scarring behavior around biomaterials placed in culture. To recreate the microglial migration and attachment to the wire, the astrocyte upregulation of GFAP and glial scar, and the neuronal isolation from the recording surface, we need to have microglia, astrocytes, and neurons in the culture. A culture without microglia could survive for longer periods than the 3 weeks we can maintain neuronal survival, and a culture without neurons could be isolated from postnatal or even adult animals, yet critical cell types would not be present in these situations. To our knowledge, ours is the first and only demonstration of this scarring behavior to be published *in vitro*, due in part to the specific presence of all three cell types known to participate in the brain's response to implanted foreign materials. The system certainly has its drawbacks, whether it is the use of embryonic cells, the short-lived viability of the cells, or the noncortical cell source, yet no other system has produced *in vitro* behavior that so closely mimics the characteristics of the *in vivo* response. We have begun to look at E17-E18 cortical cultures that also contain a relevant mix of neurons, astrocytes, and microglia, and initial results suggest a similar glial scarring response is present in these cultures as well (unpublished results).

Another multicell culture employed in our lab is the P-1 mixed glia culture. Primary mixed glia cultures are prepared from the whole brains of 1-day-old Fisher 344 rat pups or mice. After removing meninges and blood vessels, the brain tissue (minus olfactory bulbs and cerebellum) is gently triturated and 0.5×10^6 cells/well are seeded in a 24 well plate, or 0.5×10^5 cells/well in a 96 well plate. The culture medium consists of DMEM-F12 supplemented with 10% heat-inactivated FBS, 2 mM L-glutamine, 1 mM sodium pyruvate, 100 1 M nonessential amino acids, 50 U/ml penicillin, and 50 lg/ml streptomycin. Cultures are maintained at 37°C in a humidified atmosphere of 5% CO_2 and 95% air. Three days after initial seeding, cultures are replenished with 0.5 ml/well fresh medium (24 well plate) or 0.1ml/well (96 well plate). At 7 days postseeding, cultures are treated. Upon treatment, cultures are switched to fresh medium containing 2% heat-inactivated FBS), 2 mM L-glutamine, 1 mM sodium pyruvate, 50 U/ml penicillin, and 50 µg/ml streptomycin.

The P-1 mixed glia system could be a useful tool to study glial responses to foreign materials even without the presence of neurons.

4.2.3 Brain Slices

4.2.3.1 Brain Slices: Complexity Approaching *In Vivo*

Another way to model the brain environment is to use organotypic brain slice cultures. In contrast to primary cultures, these slices, as their name suggests, provide a three-dimensional representation of the cellular environment and preserve cell-to-cell interactions. The preparation of these slices minimally raises the basal level of activation and thereby allows for a more sensitive model. In addition, fewer animals are required to create these cultures since many slices can be obtained from one brain and each slice can be maintained for months [48,49]. Finally, slices of mature animals, up to postnatal day 21 to 23, can be grown in culture, whereas primary cultures can only be produced from embryonic or neonatal animals [8]. The drawbacks to increasing the complexity are that the culture approaches the uncontrolled nature of the *in vivo* environment. Furthermore, while the slices are well suited for generating acute and chronic recordings of neurons in preserved circuits, the staining and fixation protocols necessary to perform immunocytochemical analysis on the cellular response to biomaterials is more difficult than in two-dimensional cultures or even *in vivo*, where tissue sections are easily cut and stained. The thin slices attach to growth membrane in tissue culture and are difficult to fix, section, and stain.

The complexity of these models presents its own challenges. Preserving the cellular environment eliminates the ability to produce uniformly dissociated cultures and thus increases the difficulty of attributing an observed response to a specific cell type. To assess the overall response of a particular agent, the selective vulnerability of cells in different regions of the brain, which may not be readily identifiable, must be taken into account. Furthermore, the presence of extensive networks impedes examining and tracking of individual cells. Primary dissociated cultures, in contrast, allow for easy perfusion of agents without problems of tissue absorption or interaction from neighboring cells or tissues. Finally, many studies have shown that slice cultures have not consistently replicated *in vivo* responses [50]. Thus, the slice models may provide a promising alternative or addition to current primary models, but the individual challenges they pose warrant caution in their use.

In a typical isolation procedure, the tissue slice is placed onto a semiporous membrane that is attached to a removable insert, and the assembly is placed into a media-filled well. The membrane prevents direct contact of the media to the tissue. A diagram of the basic components and an image of the interface setup are shown in Figure 4.8. Typically, slices from the brains of postnatal animals, in particular rats and mice, are used. Multiple slices of uniform thickness can be prepared at once by using a vibrating microtome or a less complex manual tissue chopper. Fresh slices of approximately 300 to 400 μm thick are cultured for 2 to 3 weeks to allow cells to stabilize before treatment with chemical or mechanical injuries. The slices can be observed for many weeks afterwards.

Although few in the neuroprosthetics community have used brain slices to study biocompatibility, the literature abounds with studies to support the use of

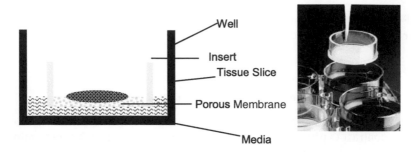

FIGURE 4.8 Diagram and image (millipore) of the Stoppini method for culturing tissue slices. The tissue is placed on an insert and positioned in a media-containing well without direct contact by the tissue to the media.

slices as culture models. The use of multiple techniques to characterize the behavior and morphology of cells has provided documentation for numerous *in vivo* events, including immediate and long-term responses, activated and resting state transitions, and cellular reactions to mechanical, excitotoxic, and bacterial injuries. In other studies, slices have been cocultured with dissociated cells to examine the interactions between specific cell types. The versatility that slice cultures offer indicates that they may be used to isolate and study a broad range of disease mechanisms.

4.2.3.2 Brain Slices: Biocompatibility Studies

Studies using this model to assess the biocompatibility of implants and prosthetics are limited. Within the field of biocompatibility, only a handful of studies have examined the response to neural electrodes. Koeneman and colleagues conducted one of the first studies, embedding single polybenzylcyclobutene implants in rat brain slices and studying the tissue response over a 2-week period [48]. Using cell-specific markers to simultaneously visualize the number, morphology, and localization of different cell types near the implant, they showed that glia and neurons made extensive contact with the implants and, further, that the implants did not promote cell death (Figure 4.9). These observations suggested that the electrodes were biocompatible. Although quantitative assays are needed to support their conclusions, the study showed that brain slice models could be used as an alternative to assess neural electrode biocompatibility and provided techniques in which such assessments could be performed.

Using a different technique, Kristensen et al. assessed neural implant biocompatibility by growing hippocampal slices directly on microelectrode arrays and analyzing the structural interaction between the array and native cells [51]. Although they observed formation of a glial scar at the base of the electrode, though not at the tip, the histological patterns in slices grown on arrays were comparable to those grown on standard membranes. Furthermore, slices grown on arrays did not become more susceptible to excitotoxins and neurotoxins. These results, in addition to the possibility that the scar may have resulted from initial tissue injury, led them to conclude that the arrays are biocompatible with tissue slices.

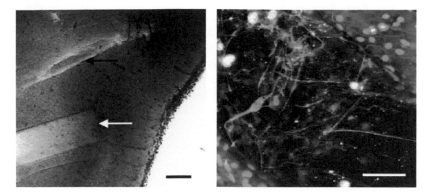

FIGURE 4.9 (Left) Electrode implanted in a 300-μm coronal slice: black arrow = insertion only, white = implant. (Right) Confocal image of the electrode after 7 days: blue = nuclei, orange = GFAP, green = neurons. (Adapted from Koeneman, B. A., Lee, K. K., Singh, A., He, J. P., Raupp, G. B., Panitch, A., and Capco, D. G., *J. Neurosci. Methods* 137, 257–263, 2004. With permission.) **(See color insert.)**

Bypassing traditional methods of slice culture that maintain neuronal elecro-physiological activity, Bjornsson et al. recently used 500-μm thick cortical slices to measure the effects of electrode insertion parameters on tissue strain [52]. In this case, the experimenters needed the three-dimensional structure and vascular features that are unavailable in cell culture but needed an *ex vivo* setup to conduct quantitative analysis on the tissue response to insertion. Their analysis found that a faster insertion results in lower tissue deformation, although large variability was found between insertions, and this variability was heavily influenced by cortical surface features such as vascular elements.

4.3 CONCLUSIONS

In vivo studies have always been, and will continue to be, the gold standard in evaluating device–tissue interactions. However, *in vivo* studies are costly, time consuming, and often cannot provide the mechanistic insight necessary to understand the underlying problems causing device failure. Without the cellular and molecular mechanisms that controlled *in vitro* cell culture experiments can provide, the quest for a nonfouling, nonscarring electrode design may continue to progress through the slow, seemingly random path that subcutaneous devices such as implantable glucose sensors have taken. However, the variety and richness of the *in vitro* cell culture models available for the brain may accelerate or help guide the *in vivo* experiments and novel electrode designs. Simple cell line experiments that cost on the order of several hundreds of dollars can prevent costly problems of *in vivo* material cytotoxicity, while primary cell cultures can help to explain why one material attracts neurite growth when another repels neuronal attachment and elongation. Cell cultures involving multiple interacting cell types or even slices of brain tissue can reproduce glial scarring and electrode failure under controlled, observable conditions and potentially eliminate the need for *in vivo* experiments except as a validation of *in vitro* results. At all times with *in vitro* experimentation, one must keep in mind that cell cultures are an imperfect model of *in vivo* behavior and that as the *in vitro* system becomes more rigid and controlled, it moves away from the uncontrolled, complicated *in vivo* environment the model is emulating. However, by keeping the limitations of *in vitro* analysis in mind, and verifying *in vitro* findings *in vivo*, one can make great strides in understanding the tissue reaction to implanted electrodes and in the design of biocompatible devices by utilizing an appropriately chosen *in vitro* cell culture system.

REFERENCES

1. Freshney, I. Application of cell cultures to toxicology. *Cell. Biol. Toxicol.* 17, 213–230, 2001.
2. Gao, H. M., Jiang, J., Wilson, B., Zhang, W., Hong, J. S., and Liu, B. Microglial activation-mediated delayed and progressive degeneration of rat nigral dopaminergic neurons: relevance to Parkinson's disease. *J. Neurochem.* 81, 1285–1297, 2002.
3. Mcgeer, P. L., Itagaki, S., Tago, H., and Mcgeer, E. G. Expression of Hla-Dr and Interleukin-2 receptor on reactive microglia in senile dementia of the Alzheimers type. *J. Neuroimmunol.* 16, 122, 1987.
4. Mcgeer, P. L., Itagaki, S., Boyes, B. E., and Mcgeer, E. G. Reactive microglia are positive for Hla-Dr in the substantia nigra of Parkinsons and Alzheimers-disease brains. *Neurology* 38, 1285–1291, 1988.
5. Misgeld, T. and Kerschensteiner, M. *In vivo* imaging of the diseased nervous system. *Nat. Rev, Neurosci.* 7, 449–463, 2006.
6. Mcgeer, P. L., Schwab, C., Parent, A., and Doudet, D. Presence of reactive microglia in monkey substantia nigra years after 1-methyl-4-phenyl-1,2,3,6-tetrahydropyridine administration. *Ann. Neurol.* 54, 599–604, 2003.
7. Gao, H. M., Liu, B., Zhang, W. Q., and Hong, J. S. Critical role of microglial NADPH oxidase-derived free radicals in the *in vitro* MPTP model of Parkinson's disease. *FASEB J.* 17, 1954–1956, 2003.
8. Fedoroff, S. and Richardson, A. *Protocols for Neural Cell Culture*, 3rd ed. Humana Press, Totowa, NJ, 2001.
9. Lee, K. K., He, J. P., Singh, A., Massia, S., Ehteshami, G., Kim, B., and Raupp, G. Polyimide-based intracortical neural implant with improved structural stiffness, *J. Micromech. Microeng.*, 14, 32–37, 2004.
10. Singh, A., Ehteshami, G., Massia, S., He, J. P., Storer, R. G., and Raupp, G. Glial cell and fibroblast cytotoxicity study on plasma-deposited diamond-like carbon coatings. *Biomaterials* 24, 5083–5089, 2003.
11. Massia, S. P., Holecko, M. M., and Ehteshami, G. R. *In vitro* assessment of bioactive coatings for neural implant applications. *J. Biomed. Mater. Res. A* 68A, 177–186, 2004.
12. Kam, L., Shain, W., Turner, J. N., and Bizios, R. Selective adhesion of astrocytes to surfaces modified with immobilized peptides. *Biomaterials* 23, 511–515, 2002.
13. Moxon, K. A., Kalkhoran, N. M., Markert, M., Sambito, M. A., McKenzie, J. L., and Webster, J. T. Nanostructured surface modification of ceramic-based microelectrodes to enhance biocompatibility for a direct brain-machine interface. *IEEE Trans. Biomed. Eng.* 51, 881–889, 2004.
14. Cui, X. Y., Lee, V. A., Raphael, Y., Wiler, J. A., Hetke, J. F., Anderson, D. J., and Martin, D. C. Surface modification of neural recording electrodes with conducting polymer/biomolecule blends. *J. Biomed. Mater. Res.* 56, 261–272, 2001.

15. Martin, D. L. and Shain, W. High-affinity transport of taurine and beta-alanine and low affinity transport of gamma-aminobutyric acid by a single transport-system in cultured glioma-cells. *J. Biol. Chem.* 254, 7076–7084, 1979.
16. Kam, L., Shain, W., Turner, J. N., and Bizios, R. Correlation of astroglial cell function on micro-patterned surfaces with specific geometric parameters. *Biomaterials* 20, 2343–2350, 1999.
17. Turner, A. M. P., Dowell, N., Turner, S. W. P., Kam, L., Isaacson, M., Turner, J. N., Craighead, H. G., and Shain, W. Attachment of astroglial cells to microfabricated pillar arrays of different geometries. *J. Biomed. Mater. Res.* 51, 430–441, 2000.
18. St. John, P. M., Kam, L., Turner, S. W., Craighead, H. G., Issacson, M., Turner, J. N., and Shain, W. Preferential glial cell attachment to microcontact printed surfaces. *J. Neurosci. Methods* 75, 171–177, 1997.
19. Craighead, H., Turner, S. W., Davis, R. C., James, C. D., Perez, A. M., St. John, P. M., Isaacson, M., Kam, L., Shain, W., Turner, J. N., and Banker, G. Chemical and topographical surface modification for control of central nervous sytem cell adhesion. *J. Biomed. Microdevices* 1, 49–64, 1998.
20. Tzeng, S. F., Huang, H. Y. Downregulation of inducible nitric oxide synthetase by neurotrophin-3 in microglia. *J. Cell. Biochem.* 90, 227–233, 2003.
21. Kremlev, S. G., Roberts, R. L., and Palmer, C. Differential expression of chemokines and chemokine receptors during microglial activation and inhibition. *J. Neuroimmunol.* 149, 1–9, 2004.
22. Wang, T. G., Liu, B., Zhang, W., Wilson, B., and Hong, J. S. Andrographolide reduces inflammation-mediated dopaminergic neurodegeneration in mesencephalic neuron-glia cultures by inhibiting microglial activation. *J. Pharmacol. Exp. Ther.* 308, 975–983, 2004.
23. Pei, Z., Pang, H., Qian, L., Yang, S. N., Wang, T. G., Zhang, W., Wu, X. F., Dallas, S., Wilson, B., Reece, J. M., Miller, D. S., Hong, J. S., and Block, M. L. MAC1 mediates LPS-induced superoxide from microglia: The role of phagocytosis receptors dopaminergic neurotoxicity. *Glia* 58, 1362–1373, 2007.
24. Dobbertin, A., Rhodes, K. E., Garwood, J., Properzi, F., Heck, N., Rogers, J. H., Fawcett, J. W., and Faissner, A. Regulation of RPTP beta/phosphacan expression and glycosaminoglycan epitopes in injured brain and cytokine-treated glia. *Mol. Cell. Neurosci.* 24, 951–971, 2003.
25. Ravenscroft, M. S., Bateman, K. E., Shaffer, K. M., Schessler, H. M., Jung, D. R., Schneider, T. W., Montgomery, C. B., Custer, T. L., Schaffner, A. E., Liu, Q. Y., Li, Y. X., Barker, J. L., and Hickman, J. J. Developmental neurobiology implications from fabrication and analysis of hippocampal neuronal networks on patterned silane-modified surfaces. *J. Am. Chem. Soc.* 120, 12169–12177, 1998.
26. Takano, M., Horie, M., Narahara, M., Miyake, M., and Okamoto, H. Expression of kininogen mRNAs and plasma kallikrein mRNA by cultured neurons, astrocytes and meningeal cells in the rat brain. *Immunopharmacology* 45, 121–126, 1999.
27. Potter, S. M. and DeMarse, T. B. A new approach to neural cell culture for long-term studies. *J. Neurosci. Methods* 110, 17–24, 2001.
28. McCarthy, K. D. and De Vellis, J. Preparation of separate astroglial and oligodendroglial cell-cultures from rat cerebral tissue. *J. Cell Biol.* 85, 890–902, 1980.
29. Liu, B., Du, L. N., and Hong, J. S. Naloxone protects rat dopaminergic neurons against inflammatory damage through inhibition of microglia activation and superoxide generation. *J. Pharmacol. Exp. Ther.* 293, 607–617, 2000.

30. Basu, A., Krady, J. K., Enterline, J. R., and Levison, S. W. Transforming growth factor beta 1 prevents IL-1 beta-induced microglial activation, whereas TNF alpha- and IL-6-stimulated activation are not antagonized. *Glia* 40, 109–120, 2002.

31. Passaquin, A. C., Schreier, W. A., and De Vellis, J. Gene-expression in astrocytes is affected by subculture. *Int. J. Dev. Neurosci.* 12, 363–372, 1994.

32. Wu, X. F., Chen, P. S., Dallas, S., Wilson, B., Block, M. L., Wang, C. C., Kinyamu, H., Lu, N., Gao, X., Leng, Y., Chuang, D. M., Zhang, W., Zhao, J., and Hong, J. S. Histone deacetylase (HDAC) inhibitors are neurotrophic and protective on dopaminergic neurons: role of histone acetylation on BDNF and GDNF gene transcription in astrocytes. In review *Molecular Psychiatry*, 2008.

33. Biran, R., Noble, M. D., and Tresco, P. A. Characterization of cortical astrocytes on materials of differing surface chemistry. *J. Biomed. Mater. Res.*, 46, 150–159, 1999.

34. Guenard, V., Frisch, G., and Wood, P . M. Effects of axonal injury on astrocyte proliferation and morphology in vitro: Implications for astrogliosis. *Exp. Neurol.* 137, 175–190, 1996.

35. Hirsch, S. and Bahr, M. Immunocytochemical characterization of reactive optic nerve astrocytes and meningeal cells. *Glia* 26, 36–46, 1999.

36. Biran, R., Noble, M. D., and Tresco, P. A. Directed nerve outgrowth is enhanced by engineered glial substrates. *Exp. Neurol.* 184, 141–152, 2003.

37. Polikov, V. S., Block, M. L., Fellous, J. M., Hong, J. S., and Reichert, W. M. *In vitro* model of glial scarring around neuroelectrodes chronically implanted in the CNS. *Biomaterials* 27, 5368–5376, 2006.

38. Li, G. R., Cui, G., Tzeng, N. S., Wei, S. J., Wang, T. G., Block, M. L., and Hong, J. S. Femtomolar concentrations of dextromethorphan protect mesencephalic dopaminergic neurons from inflammatory damage. *FASEB J.* 19, 489–496, 2005.

39. Wang, T. G., Pei, Z., Zhang, W., Liu, B., Langenbach, R., Lee, C., Wilson, B., Reece, J. M., Miller, D. S., and Hong, J. S. MPP+-induced COX-2 activation and subsequent dopaminergic neurodegeneration. *FASEB J.* 19, 1134–1136, 2005.

40. Wang, T. G., Liu, B., Qin, L. Y., Wilson, B., and Hong, J. S. Protective effect of the SOD/catalase mimetic MnTMPyP on inflammation-mediated dopaminergic neurodegeneration in mesencephalic neuronal-glial cultures. *J. Neuroimmunol.* 147, 68–72, 2004.

41. Qin, L. Y., Li, G. R., Qian, X., Liu, Y. X., Wu, X. F., Liu, B., Hong, J. S., and Block, M. L. Interactive role of the toll-like receptor 4 and reactive oxygen species in LPS-induced microglia activation. *Glia* 52, 78–84, 2005.

42. Qin, L. Y., Liu, Y. X., Wang, T. G., Wei, S. J., Block, M. L., Wilson, B., Liu, B., and Hong, J. S. NADPH oxidase mediates lipopolysaccharide-induced neurotoxicity and proinflammatory gene expression in activated microglia. *J. Biol. Chem.* 279, 1415–1421, 2004.

43. Mcmillian, M. K., Thai, L., Hong, J. S., Ocallaghan, J. P., and Pennypacker, K. R. Brain injury in a dish: A model for reactive gliosis. *Trends Neurosci.* 17, 138–142, 1994.

44. Mcmillian, M. K., Pennypacker, K. R., Thai, L., Wu, G. C., Suh, H. H., Simmons, K. L., Hudson, P. M., Sawin, S. B. M., and Hong, J. S. Dexamethasone and forskolin synergistically increase [Met(5)]enkephalin accumulation in mixed brain cell cultures. *Brain Res.* 730, 67–74, 1996.

45. Kim, W. G., Mohney, R. P., Wilson, B., Jeohn, G. H., Liu, B., and Hong, J. S. Regional difference in susceptibility to lipopolysaccharide-induced neurotoxicity in the rat brain: Role of microglia. *J. Neurosci.* 20, 6309–6316, 2000.

46. Liu, B., Jiang, J. W., Wilson, B. C., Du, L., Yang, S. N., Wang, J. Y., Wu, G. C., Cao, X. D., and Hong, J. S. Systemic infusion of naloxone reduces degeneration of rat substantia nigral dopaminergic neurons induced by intranigral injection of lipopolysaccharide. *J. Pharmacol. Exp. Ther.* 295, 125–132, 2000.

47. Li, G. R., Liu, Y. X., Tzeng, N. S., Cui, G., Block, M. L., Wilson, B., Qin, L. Y., Wang, T. G., Liu, B., Liu, J., and Hong, J. S. Protective effect of dextromethorphan against endotoxic shock in mice. *Biochem. Pharmacol.* 69, 233–240, 2005.

48. Koeneman, B. A., Lee, K. K., Singh, A., He, J. P., Raupp, G. B., Panitch, A., and Capco, D. G. An *ex vivo* method for evaluating the biocompatibility of neural electrodes in rat brain slice cultures. *J. Neurosci. Methods* 137, 257–263, 2004.

49. Kunkler, P. E. and Kraig, R. P. Reactive astrocytosis from excitotoxic injury in hippocampal organ culture parallels that seen *in vivo. J. Cereb. Blood Flow Metab.* 17, 26–43, 1997.

50. Noraberg, J., Poulsen, F. R., Blaabjerg, M., Kristensen ,B. W., Bonde, C., Montero, M., Meyer, M., Gramsbergen, J. B., and Zimmer, J. Organotypic hippocampal slice cultures for studies of brain damage, neuroprotection and neurorepair. *Curr. Drug Ther. CNS Neurol. Disord.* 4, 435–452, 2005.

51. Kristensen, B. W., Noraberg, J., Thiebaud, P., Koudelka-Hep, M., and Zimmer, J. Biocompatibility of silicon-based arrays of electrodes coupled to organotypic hippocampal brain slice cultures. *Brain Res.* 896, 1–17, 2001.

52. Bjornsson, C. S., Oh, S. J., Al Kofahi, Y. A., Lim, Y. J., Smith, K. L., Turner, J. N., De, S., Roysam, B., Shain, W., and Kim, S. J. Effects of insertion conditions on tissue strain and vascular damage during neuroprosthetic device insertion. *J. Neural Eng.* 3, 196–207, 2006.

5 *In Vivo* Solute Diffusivity in Brain Tissue Surrounding Indwelling Neural Implants

Michael J. Bridge and Patrick A. Tresco

CONTENTS

5.1   INTRODUCTION

In this chapter we describe an analytical approach that can be used to study solute diffusivity as a function of the tissue composition surrounding a device implanted in normal, aged, damaged, or diseased brain. The method is capable of resolving changes in solute diffusivity and cellular composition at the scale of a few microns. In this chapter we describe the key features of the method including a quantitative imaging and modeling approach that can be used to assess extracellular diffusion surrounding a model implant, and we illustrate how the approach can be used to study the influence of living cells on tissue remodeling and solute transport. Available evidence suggests that the approach may be useful in sorting out the intricacy of the foreign body response, as well as examining how soluble factors released from various types of transplanted cells affect brain tissue remodeling and regional regeneration in the central nervous system.

Our current knowledge of the central nervous system (CNS) response to implants would benefit from understanding whether solute diffusion is affected by the tissue that develops around chronic implants irrespective of whether the implant is a biomedical device or a cellular-based implant. As has been described in other chapters of this book, this so called "foreign body response" develops irrespective of the size or type of device or transplant [1–11] and has a characteristic phenotype that includes inflammatory markers, reactive gliosis, and neuronal cell loss [12–14]. To date, significant effort has been focused on describing the events that accompany the implantation of such devices, tissues, and cells into brain tissue over time with a particular emphasis directed at describing the temporal and spatial nature of the events that take place at the implant–tissue interface, and assessing the potential of the response to affect device or transplant function. By all indications, a full understanding of what needs to be done to consistently interface various devices and living transplants with a variety of potential neural targets is still a ways off. Indeed, it appears that the scientific breadth and depth has not sufficiently advanced, as has, for example, occurred with cochlear implants, so that the challenge of these newer therapeutic approaches can shift from one of lack of scientific know-how to one of engineering.

Difficulties associated with the delivery of agents to the CNS have led to a growing number and variety of interventional approaches [15,16]. Although the feasibility of focal delivery devices [5,17–20] and various neuroprosthetic applications has been established [21–24], the efficacy of various approaches is often compromised by chronic use. It has become clear that improvement of CNS implant technology will require a better understanding of the nature of the tissue that develops adjacent to implants, as well as a better understanding of how to modulate the properties of the tissue.

To date, a number of experimental techniques have been utilized to characterize solute transport in the extracellular space (ECS) of the CNS. Pioneering studies were performed through measurement of radiolabeled compounds in tissue sections at different time points following ventricle perfusion in animals [25–27]. Derivation of tissue diffusivity properties was accomplished by employing solute diffusion models to the analysis of probe distribution. More recently, autoradiography of thin sections along with image analysis has been used to assess radiolabeled solute diffusion from implanted polymeric controlled release devices [28].

The real-time iontophoretic tetramethylammonium technique (also referred to as the tetramethyl ammonium [TMA+]-microelectrode method) has been used for *in vivo* assessment of CNS–ECS diffusivity and to assess changes associated with a number of physiological and pathological conditions [29–31]. In this method, TMA+ is introduced by iontophoresis from an electrode aligned parallel to a double-barreled TMA+ ion-sensitive electrode (ISE) set at a fixed distance away from a source electrode. Diffusion of released TMA+ is then assessed through real-time measurement of ion concentrations in tissues where the TMA+ ISM is located. Local point concentration versus time measurements coupled with appropriate diffusion models are then employed to evaluate CNS–ECS diffusivity parameters. A limitation of the real-time iontophoretic method is that only relatively small molecular weight (MW) charged molecules (e.g., TMA+ = 74 Da) can be used as diffusion probes, and two acute penetrating injuries are required.

The integrative optical imaging (IOI) technique is an *in vitro* technique previously utilized to characterize macromolecule diffusion in CNS tissue [32–34]. A bolus of fluorescently labeled solute is released into a thick (~400 μm) brain section mounted under a standard upright microscope. Diffusion is then monitored and measured by capturing fluorescence images at increasing longer time intervals. Theoretical expressions are then employed to account for both in-focus and out-of-focus fluorescence signals. Such measurements yield a series of probe distribution profiles for the time points collected, from which diffusivity is evaluated using an exponential model of solute diffusion. Based on methods similar to the IOI technique, more recently multiphoton microscopy has been employed to evaluate the distribution of labeled nerve growth factor and dextran probes in brain slices [35]. The investigators suggest that the technique is an improvement since out-of-focus fluorescence does not blur recorded images, thereby simplifying interpretation of probe distributions.

Another recently developed technique referred to as "dual-probe" microdialysis has been employed to investigate CNS–ECS diffusivity parameters in rat brains [36,37]. A microdialysis probe is used to infuse a radiotracer into the rat brain, and another microdialysis probe, positioned in proximity to the first (~1 mm), is then utilized to sample changes in the local solute concentration over time. Fitting of local point concentration versus time using appropriate mathematical models then enables derivation of bulk CNS–ECS diffusivity parameters for the interstitial region located between the probes.

Recently we described a novel tissue access device (TAD) based on hollow fiber membrane (HFM) brain tissue implants that were used to examine whether cells transplanted into the TAD influenced solute diffusion in the brain tissue that surrounds the implant [38–40]. Unlike other methods, TADs assess how various tissue components affect the transport of soluble factors after a chronic intervention. In this case, the device can be implanted prior to introducing the cells or the solute molecules, which allows one to design experiments that assesses the foreign body response at different time points and to examine what might be expected following the transplantation of cells after a long or short indwelling times. Also, since the probe can be introduced into the lumen of the device through a skin incision without requiring device removal, transport can be examined through native tissue that is not disturbed by the acute damage of a penetrating ion-selective microelectrode or other acute trauma that is required with the previously described methods.

This chapter presents undescribed details of how the TAD and finite element modeling can be used to evaluate solute distribution in brain tissue by estimating the diffusivity properties of the surrounding tissue at 2 μm spatial intervals. The diffusivities are related to specific populations of cells adjacent to the implant using cell–type-specific markers. We also validate the methodology using a gel solute diffusion model system and then describe how the method can be used to examine whether natural products constitutively expressed and released from transplanted cells influence the tissue properties surrounding the implantation site.

5.2 METHODS

5.2.1 DEVICE AND IMPLANTATION PROCEDURE

A schematic of the cell encapsulation device is shown in Figure 5.1A. The implants were fabricated and sterilized as described previously [38–40]. A photograph of a representative device is shown in Figure 5.1F. A poly(acrylonitrile-vinyl chloride) (PAN-PVC) HFM was fabricated as described [41] with an asymmetric morphology with an inner diameter of ~700 μm and wall thickness of ~90 μm (see Figure 5.1E). The 70-kDa dextran diffusion coefficient for the naïve membrane used was ~6.0 × $10^{-8} cm^2/sec$ [42].

Cell coil constructs were loaded with cells and implanted using stereotaxic methods as previously described [38–40]. Collagen-entrapped meningeal cells or collagen matrix alone were transferred into the devices prior to implantation. Cellular constructs can be prepared using a variety of cell types including anchorage- and nonanchorage-dependent cell types using a density of 3.0×10^4 entrapped cells per HFM construct. Prior to implantation loaded coil constructs are maintained in serum-free culture media so as not to induce reactivity not associated with the implanted cells.

The orientation of the device implanted within the rat brain is depicted in Figure 5.1B. A photograph of a TAD implanted in the rat brain is shown in Figure 5.1G. Typically 5 or 6 animals are utilized for both the control group (i.e., collagen only) and experimental groups (i.e., collagen-entrapped meningeal fibroblasts). Once the device is implanted and anchored to the cranium, the cell coil element is inserted into the HFM chamber (Figure 5.1C), the access cap is snapped in place, and the overlying dermis is closed with sutures. Alternatively, the cells or a sustained-release polymer can be placed in the implanted device at a later time point to study the influence of the cells or drugs on tissue remodeling.

5.2.1.1 Diffusion Experiments

Following a recovery period animals are reanesthetized, the dermis overlying the implanted TAD is opened, the access cap is removed, and the cell coil construct is withdrawn from the device. With the aid of a low-flow microsyringe pump, a lysine-

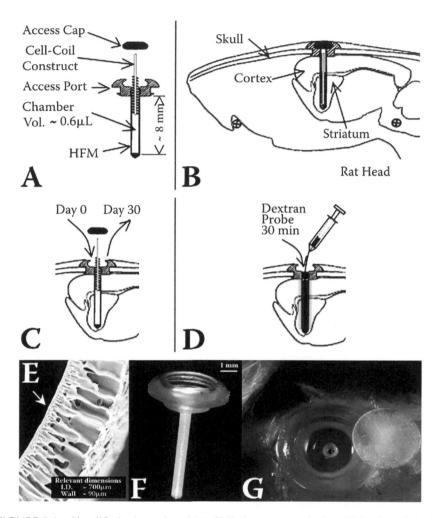

FIGURE 5.1 Simplified schematics of (A) CNS tissue access device; (B) implanted device orientation in rat brain; (C) insertion of cell-coil construct; and (D) loading and extraction of probe solution into and from the intralumenal chamber of the device for CNS diffusivity experiment runs. (E) Scanning electron micrograph of PAN-PVC hollow fiber membrane employed in assembly of TAD. The arrow points to the intralumenal permselective skin layer of the HFM. (F) Photograph of actual TAD employed in the study. (G) Photograph of TAD implanted in rat brain with access cap removed.

fixable Texas Red (TR)-labeled 70-kDa dextran solute probe (Molecular Probes, Eugene, Oregon) in phosphate buffered saline (PBS) at ~0.25 mg/ml is then transferred into the TAD chambers up to the concave region of the access port originally occupied by the access cap. The probe solution is then allowed to diffuse into the surrounding tissue for a 30-minute period (Figure 5.1D), after which the chamber is flushed and loaded with a 4% para-0.25% glutaraldehyde in 1xPBS tissue fixative. Animals are then immediately perfused transcardially with the same tissue fixative mixture. Brains are explanted, placed in tissue fixative overnight at 3°C, and maintained in a PBS solution with 0.1% sodium azide (NaA) at 3°C.

5.2.1.2   Histological Processing of Tissue Samples

A vibratome (Technical Products International, St. Louis, Missouri) is utilized to make horizontal tissue sections across the retrieved brain samples. Slices at 20 to 40 µm are serially collected and stored in 12-well plates containing a 1×PBS solution with 0.1% NaA at 3°C. Throughout sectioning and subsequent processing of sections, care is taken to minimize light exposure of TR-dextran-laden tissue sections.

Batch processing of sections is utilized for all subsequent staining procedures to limit staining variability and to ensure consistent treatment of all samples. Incubation of sections with cell-specific markers allows regional identification of particular cell types, as will be illustrated here using markers for reactive astrocytes and macrophages and reactive microglia. A first set of tissue sections, consisting of every fourth serial section (200 µm spacing), is stained for glial fibrillary acidic protein (GFAP), an intermediate filament protein known to be upregulated by reactive astrocytes comprising glial scars [43]. Sections are exposed to a 4% goat serum in 1×PBS with 0.5% triton to permeablize cell membranes for 1H at room temperature (~22°C). Primary antibody consisting of appropriate antisera, for instance, polyclonal rabbit anti-GFAP (1:2000 dilution, mouse IgG, Dako) is applied overnight at 4°C. An appropriate secondary antibody such as Alexa 488–labeled goat anti-rabbit IgG (H+L) (1:220 dilution, Molecular Probes) is then applied for a minimum of 1 H at ~22°C. Additional sections serve as primary controls through incubation with only the applied secondary antibody.

A second set of tissue sections with 400-µm spacing (every eighth section) is stained for another marker, for instance, ED1, an antigen marker commonly employed to discriminate between activated microglia and macrophages in the CNS [44]. With the exception of exposure to triton during the blocking step, a similar immunostaining protocol is used for ED1 detection. The primary antibody consists of mouse anti-ED1 (1:1000 dilution, rat IgG1, Serotec), and the secondary is Alexa 488–labeled goat anti-mouse IgG1 (1:220 dilution, Molecular Probes). ED1 staining primary control sections are also prepared similar to the previous protocol.

All sections are then counterstained with the nuclear dye 4',6-diamidino-2-phenylindole, dihydrochloride (DAPI, Molecular Probes), mounted on standard microscope slides using Fluoromount-G (Molecular Probes), and stored in the dark at 3°C. Additional sections, to be utilized to assess the relative concentrations of TR-Dextran probe across the transmembrane region, are also mounted on slides.

5.2.1.3   Fluorescence Imaging of Samples

An epifluorescent microscope (Nikon Eclipse E600 upright microscope configured with appropriate filter cube sets and a Coolsnap camera for digital image capturing) was used to capture images of the TR-labeled 70-kDa dextran, Alexa 488-labeled GFAP or ED1 antigens, and DAPI nuclear stains. Four 20× magnification images were collected for each section around the circumference of the membrane specifically at rostral, lateral, caudal, and medial anatomical orientations. The capture times utilized for fluorescence imaging of solute probe and stains is constant for all samples processed.

The raw images undergo dark field subtraction and flat field light correction using Image-Pro software. Dark field subtraction is applied to composite images by subtraction of a dark field calibration image generated through averaging of 5 images acquired under no excitation light conditions. Flat-field correction is then applied to composite images to correct for uneven illumination of samples in the acquisition region. This correction is based on images captured of calibration slides containing homogenous distributions of the fluorescent probes used in the study. Flat-field correction is applied to experimental sample images using the following relationship:

$$(I_{exp} / I_{flat}) \times S \tag{5.1}$$

where I_{exp} and I_{flat} represent the experimental and flat-field calibration images, and S represents a pixel intensity scale factor derived from peak intensity values appearing in calibration images. Images are then stored as tagged image file formats (TIFFs), which maintain intensity data as discrete channel information.

5.2.1.4   Quantitative Image Analysis

To ensure nonbiased fluorescence pixel intensity sampling of collected images, a Labview (National Instruments Co., Austin, Texas)–based image analysis program was created, and qualified in preliminary studies, specifically for this experimental application (introduced in [40]). In the analysis program created (see Figure 5.2) three points were utilized to delineate a reference arc at the outer wall of the HFM. The angular location, radius, and size of the reference arc is adjustable, enabling coordination of the radial axis between sections as well as exclusion of processing defects that may be present in images such as air bubbles. Once a reference arc is delineated, quantitative fluorescence intensity sampling profiles (QFPs) are generated extending normal to the arc at a density of ~1.7 degree/profile, and on average for the present study yielded ~24 sampling profiles per image analyzed. The QFPs generated are radially indexed in ~0.5-μm increments that originate 100 μm back from the delineated arc to enable analysis of the transmembrane regions and range from 500 to 700 μm in total length. At each radial increment an antialias pixel extraction algorithm is applied to derive an intensity value based on a weighted averaging of pixels in nearest proximity (as depicted in inset of Figure 5.2).

The composite QFP for the image is then calculated by averaging all coinciding increments of the sampling profiles generated and is performed simultaneously for all color channels present in images. When profile intensity averaging is performed on solute diffusion probe distributions, values falling outside of two standard deviations are discarded and the average recalculated. This filter is not applied to histological stains since punctate patterns are expected, given that the populations of cell types of interest are discrete regional entities likely to vary in the circumferential

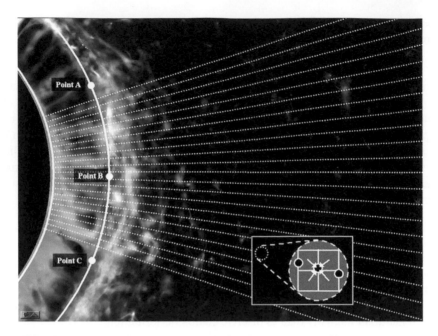

FIGURE 5.2 Demonstration of the quantitative fluorescence profiling sampling method employed in study. Three points (A, B, and C) are used to delineate a reference arc at the outer surface of the HFM. Radially defined sampling arrays are then generated at regularly spaced intervals. As depicted in the inset figure, multiple pixels in the immediate vicinity of each sampling array point are used in the evaluation of its intensity value.

orientation of samples. The presence of tissue section processing irregularities, such as partial tears in the tissue, displacement of tissue from the membrane interface, or air bubble entrapment within the capture region, results in the rejection of all QFPs generated for that image. However, only rarely do all four orientations captured for a specific tissue section have to be excluded from analyses.

5.2.1.5 Data Processing

Following QFP analysis the mean intensity profiles (i.e., mean of composite QFPs) are calculated for the stains and the 70-kDa TR-dextran probe. To specifically map and discriminate ED1 and GFAP cell marker spatial distributions, composite QFPs are normalized prior to calculation of mean intensity profiles. Marker profiles are normalized to peak relative fluorescence intensity values following subtraction of the appropriate background profiles derived from QFP analysis of primary control sections.

The 70-kDa TR-dextran composite QFP for the dextran probe is not normalized prior to mean calculations. The background subtraction applied is based on 10 images captured of the contralateral hemisphere of randomly selected brain sections from both study groups. Since saturated intensity values are typically found for the dextran probe within the membrane region, only the portion representing the adjacent brain tissue region is considered in later analyses. Following mean profile calculation dextran curves are normalized through division with the intensity value at the radius representing the membrane–tissue (m–t) interface.

5.2.1.6 Gel Diffusivity Characterization Experiments

Gel medium utilized for gel diffusivity characterization experiments consisted of a 40:60 ratio of porcine skin gelatin to high-temperature-melt agarose at a 1.5% weight-to-volume (w/v) content. Gel solutions were prepared through addition of gelatin and agarose components to a 1×PBS solution with 0.1% NaA. A 12-well plate served as the mold for the embedding process, with TADs maintained radially centered and normal to the bottom surface of wells during solidification of the heated gel solution. Wells were filled with the gel solution up to the midpoint of the access port component of the TAD. Gel-embedded TAD samples were then removed from molds and placed in 6 well plates containing 1×PBS solution with 0.1% NaA to ensure their complete hydration prior to diffusion runs.

The protocol used for *in vitro* gel diffusion experiments closely mimicked that used for *in vivo* experiments. The level of PBS solution in wells was brought even to that of the top surface of gels. The PBS solution in the intralumenal chamber of the TAD was then exchanged with an Alexa 488–labeled streptavidin solute probe (MW ~54 kDa, Molecular Probes) in PBS at a concentration of ~0.25 mg/ml. The labeled protein was then allowed to diffuse into the surrounding gel for 10- ($n = 2$), 20- ($n = 2$), 30- ($n = 1$), 60- ($n = 1$), 120- ($n = 1$), or 180-minute ($n = 1$) periods. At the end of each diffusion run the chamber was flushed with PBS and loaded with 10% glutaraldehyde/PBS fixative solution. The surrounding buffer solution was also replaced with the fixative solution. Samples were then stored at 4°C for a minimum of 12 h prior to sectioning.

Following fixation the fragility of gels was such that sectioning with TADs in place could not be performed. Following removal of TADs the gels were carefully cross-sectioned at the axial midlength. The original top and bottom surfaces of gel were then mounted on sectioning blocks, and a vibratome was utilized to prepare a level surface using a 150-μm cutting interval. Once a level surface was prepared, two 100-μm-thick samples were sectioned and stored in buffer solution. This process was repeated for the remaining cross sections for a total of four 100-μm sections collected per gel sample. Four sections from a gel not exposed to labeled protein were collected for background intensity determination.

Fluorescent gel images were captured in the previously described manner with the exception that 10× magnification was required, given the extent of probe diffusion. Four images, representing upper and lower left and right quadrants, were captured per section. QFP analysis of images was then performed as described above. Microfractures and irregularities present at the gel–membrane interface that resulted from TAD extraction and sectioning processes required discarding of a large number (~55 to 60%) of the composite QFPs prior to evaluation of mean intensity profiles.

5.2.1.7 Evaluation of *In Vivo* Diffusivity Properties: Central Nervous System (CNS)–Extracellular Space (ECS) Solute Diffusion Modeling

Approximation of CNS tissue diffusivity properties was accomplished by fitting the observed TR-labeled 70-kDa dextran probe mean intensity profiles with that derived from simulations employing a theoretical model describing solute diffusion in the CNS–ECS. Given that profiles represent transient-state distributions and given the expected variations in the diffusivity characteristics of the HFM intraluminal space, membrane wall space, reactive microglia–macrophage layer, glial scar layer, and the unaffected (naïve) CNS tissue regions, we elected to employ a radial finite element difference model that enables one to account for differences in their diffusivity properties. Fitting of experimental data by simulated results was accomplished through trial and error adjustments of parameters relevant to the CNS–ECS solute diffusion model. A similar approach was utilized to estimate an effective diffusivity coefficient for implanted PAN-PVC HFMs as well as for the streptavidin probe in the gel medium employed.

The mathematical model employed for CNS–ECS diffusion simulations was based on Fick's second law applied to a system in which solute diffusion occurs symmetrically in a radially outward direction:

$$\frac{\partial C(r,t)}{\partial t} = D\left(\frac{\partial^2 C(r,t)}{\partial r^2}\right) \tag{5.2}$$

where $C(r,t)$ is the solute concentration at a distance r from the origin at time t, and D is the solute diffusion coefficient. Equation (5.2) only provides a macroscopic description of solute diffusion in a homogeneous system (i.e., continuum diffusion or free solution medium) such as water. For this reason Fick's second law is commonly employed to evaluate the diffusion coefficient of a particular solute in pure water (D^w). It is a convenient benchmark for solute diffusion in heterogeneous systems such as tissue to be compared. However, on a microscopic scale solute diffusion in the CNS is restricted to the complex geometric structures that comprise the ECS pathways. The presence of solute in tissue is therefore discontinuous in nature and has historically been introduced in tissue diffusion models through incorporation of a term referred to as the void volume fraction (α). In regard to CNS–ECS solute diffusion modeling, the void volume fraction simply represents the ratio of ECS volume to overall tissue volume.

In the literature related to this subject, when an equation of continuum diffusion, such as Equation (5.2), is used in the evaluation of tissue diffusivity properties, it is common to refer to resultant values of D as apparent diffusion coefficients (D^a). In addition, investigators of CNS–ECS refer to the degree to which solute diffusion in tissue is reduced (i.e., hindered diffusion) compared to diffusion in pure water as "tortuosity" [29,45,46]. Knowledge of D^a subsequently can be used to calculate the ECS diffusivity parameter tortuosity (λ) defined as

$$\lambda = \sqrt{D^w / D^a} \tag{5.3}$$

The geometrical symmetry in the two-dimensional radial model employed for this analysis enables simplification to a one-dimensional system for CNS–ECS solute diffusion simulations. The finite element system is therefore composed of a linear nodal array of elements that originates in the center of the HFM. In finite difference form Equation (5.2) may be approximated:

$$\frac{C_{i,n+1} - C_{i,n}}{\Delta t} = \frac{D_i}{\Delta r^2} (C_{i+1,n} - 2C_{i,n} + C_{i-1,n}),$$ (5.4)

where Δr represents the distance between nodes and Δt represents the incremental time interval used for the simulation run. Note that the right-hand side of Equation (5.4) multiplied by Δt yields the incremental change in solute concentration (ΔC_i) for nodal element i. The other terms present in Equation (5.4) are defined as

$$C_{i,n} = C(r_i, t_n)$$

$$r_i = (\Delta r)i, \quad \text{for} \quad 0 \le i \le i_{norm}$$

$$t_n = (\Delta t)n, \quad \text{for} \quad 0 < n \le n_{max} \quad \text{where} \quad n_{max} = t_{end}/\Delta t \quad \text{and}$$ (5.5)

$$D_i = \begin{cases} D_{int\,ra}^w & 0 \le i \le i_{intra} \\ D_{memb} & i_{intra} < i \le i_{memb} \\ D_{micro}^a & i_{memb} < i \le i_{micro} \\ D_{scar}^a & i_{micro} < i \le i_{scar} \\ D_{norm}^a & i_{scar} < i \le i_{norm} \end{cases} \quad \text{for} \quad \begin{array}{l} i_{intra} = R_{intra}/\Delta r \\ i_{memb} = R_{memb}/\Delta r \\ i_{micro} = R_{micro}/\Delta r \\ i_{scar} = R_{scar}/\Delta r \\ i_{norm} = R_{norm}/\Delta r \end{array}$$

where R_{intra}, R_{memb}, R_{micro}, R_{scar}, and R_{norm}, respectively, delineate the radial distances from the origin to the intralumenal and extralumenal membrane surfaces and the outer tissue boundaries of the reactive microglia, glial scar, and naïve CNS tissue regions. The apparent diffusivity of scar tissue (D_{scar}^a) is defined based on attaining a reasonable fit to the observed results, which in the case of our preliminary study yielded the best fits when defined with a distance dependency (i.e., $D_{scar}^a = f(r_i)$).

Diffusion of solute occurs radially outward in the experimental two-dimensional system that the linear finite element model is intended to simulate. Therefore, nodal elements (r_i) in our linear model need to represent two-dimensional annular regions (A_i) of the system. Since the annular regions grow in size as elements move outward from the origin (i.e., $A_i > A_{i-1}$), it is necessary to incorporate this aspect into the mathematical desciption of the linear system presented in Equation (5.4). This was accomplished by reformulating Equation (5.4) to solve for elemental changes in mass (Δm_i) at each incremental time period (Δt), with the elemental concentrations now defined:

$$C_i = m_i / A_i,$$ (5.6)

Inserting Equation (5.6) into Equation (5.4) and solving for Δm_i yields

$$\Delta m_i = m_{i,n+1} - m_{i,n} = \frac{D_i}{\Delta r^2}((C_{i,n} - C_{i-1,n})A_{i-1} + (C_{i,n} - C_{i+1,n})A_i)\Delta t \qquad (5.7)$$

where

$$A_i = \alpha_i ((r_{i+1})^2 - (r_i)^2)\pi \qquad (5.8)$$

The first term on the right side of Equation (5.8) (α_i) accounts for the void volume fraction of the ECS diffusivity parameter in our diffusion model, with α_{micro}, α_{scar}, and α_{norm} defined in a manner similar to D_i in Equation (5.5), with the exception that $\alpha_i = 1$ for $i \le i_{memb}$. The elemental mass at time $t = n + 1$ is then calculated as

$$m_{i,n+1} = m_{i,n} + \Delta m_{i,n+1} \qquad (5.9)$$

The CNS–ECS solute diffusivity simulation therefore consisted of solving of Equation (5.7) in conjunction with Equation (5.9) with a new nodal mass array [m_0, m_1, ... m_{norm}] and the subsequent concentration nodal array (using Equation (5.6)) calculated at each incremental time interval. Reported simulated probe distribution profiles represent the nodal concentration array values present at the end of a simulation runs. To ensure the stability of the simulation, Δt was chosen by

$$\Delta t < \frac{\Delta r^2}{2D_i^{min}} \qquad (5.10)$$

where D_i^{min} represents the lowest value of D_i applied in the simulation.

At the onset of studies it was unclear whether the mass of probe available in the chamber should be modeled as finite or infinite, given the extra probe solution placed in the access port region of TADs during diffusion runs. For this reason simulations were conducted using two initial conditions representative of finite and infinite solute probe reservoirs. The initial conditions employed for a finite solute mass in the intralumenal space with membrane and tissue regions void of solute was

$$C_i(r_i, t_0) = \begin{cases} 100 \\ 0 \end{cases} \text{ for } \begin{array}{l} 0 \le i \le i_{intra} \\ i_{intra} < i \le i_{norm} \end{array} \qquad (5.11)$$

For infinite reservoir simulation runs, solute mass in the intralumenal space was specified:

$$C_i(r_i, t_n) = 100 \text{ for } 0 \le i \le i_{intra} \text{ and } 0 \le n \le n_{max} \qquad (5.12)$$

The infinite sink boundary condition imposed on the last element of the nodal array was

$$C_i(r_i, t_n) = 0 \quad \text{for} \quad i = i_{max} \quad \text{and} \quad 0 \le n \le n_{max} \tag{5.13}$$

The value of i_{max} used for simulations large enough that no probe was present at the preceding node at the end of simulation runs.

All CNS solute diffusion simulations utilized a pure water probe diffusivity value $D_{intra}^w = 3.72 \times 10^{-7}$ cm²/s. This free medium diffusivity value for a 70k-Da dextran solute, also used in the calculation of tortuosity values cited in this study, was an estimate based on the empirical correlation [47]: $D = 7.667 \times 10^{-5}(MW)^{-0.47752}$, where MW is the molecular weight of the dextran molecule. The membrane diffusivity value (D_{memb}) used was based upon the results of separate simulation analyses using parameters for all tissue regions based on those of naïve CNS tissue. An apparent diffusivity value of normal tissue, $D_{norm}^a = 7.5 \times 10^{-8}$ cm²/s, derived from the literature [32], and a void fraction value of normal tissue, $\alpha_{norm} = 0.21$, also derived from the literature [48], were used for initial simulations. D_{micro}^a, D_{scar}^a, α_{micro}, and α_{scar} were determined through fitting of experimentally obtained profiles. The values $\Delta r = 2\mu m$ and $\Delta t = 0.2s$ were used for *in vivo* solute diffusion simulation analyses of control and experimental group data.

For gel diffusion simulations the region outside of the membrane was assumed to have a constant diffusivity value and void fraction value of 1. In addition, a pure water diffusivity value for the labeled streptavidin (~54 kDa) of $D_{intra}^w = 7.54 \times 10^{-}$ cm²/s was used based upon the theoretical equation of free medium protein diffusion [49]: $D = A/(MW)^m$, where $A = 2.85 \times 10^{-5}$ cm²s⁻¹g¹ᐟ³mol⁻¹ᐟ³ and $m = 1/3$. A value of $D_{memb} = 1.8 \times 10^{-7}$ cm²/s was also used and was estimated based upon previous characterizations of the HFM performed using a range of dextran solute probes. The value of Δr used to simulate solute diffusion in gels was 8 µm.

5.3   RESULTS

5.3.1   GEL DIFFUSIVITY CHARACTERIZATION EXPERIMENT

Fluorescent microscopy images representing each of the time periods utilized for test runs appear in Figure 5.3. Circular voids located near the corners of images delineate the region initially occupied by HFM prior to sectioning of fixed sample gels. Characteristic of solute diffusion phenomenon, increasing the length of time of test runs resulted in serially greater probe distributions within samples.

The results of QFP analysis of captured fluorescent images is shown in Figure 5.4, where the normalized mean intensity profiles for the various time groups are plotted. The concentrations of probe observed for the 10- and 20-minute groups appear to drop off exponentially, while the longer runs appear sigmoidal. The sigmoidal shape of the latter profiles is indicative of diffusion in a system driven by a source containing a finite mass of solute.

A finite element model of solute diffusion simulation that incorporated elements spatially representative of the TAD chamber, HFM wall, and adjacent gel medium was then utilized to estimate the effective diffusivity coefficient of the streptavidin probe by fitting of the 10- and 20-minute results. Based upon the results of simulation analyses, a free medium effective diffusivity value of $D_{eff} = 6.78 \times 10^{-7}$

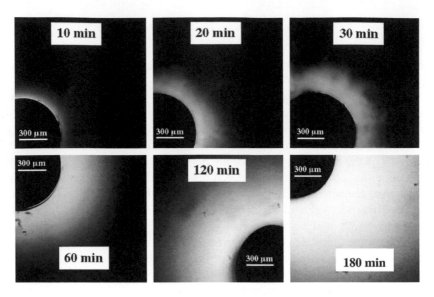

FIGURE 5.3 Representative 10× fluorescence microscopy images of Alexa 488–labeled streptavidin in 1.5% w/v 40:60 gelatin–agarose gel for experimental run times as indicated in images. The circular void regions present in images were originally occupied by the HFM that had to be extracted from gels prior to their sectioning.

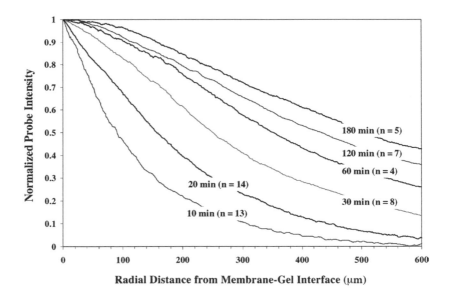

FIGURE 5.4 Normalized mean intensity profiles of Alexa 488–labeled streptavidin in 1.5% w/v 40:60 gelatin–agarose gel for experimental run times as indicated in the graph. The "n" values displayed adjacent to the profile labels represent the number of composite QFP profiles utilized in the evaluation of mean intensity profiles. The profiles were normalized based upon the mean intensity value at the surface of the gel initially adjacent to the TAD membrane (0 μm in the chart).

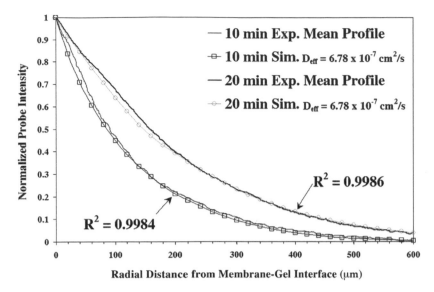

FIGURE 5.5   Alexa 488–labeled streptavidin normalized mean intensity profiles for 10-and 20-minute experimental run time sample groups along with 10- and 20-minute simulated probe distribution profiles attained from simulation runs that utilized an effective gel diffusion coefficient of 6.78×10^{-7} cm²/s to attain the displayed fits of observed experimental probe distribution profiles. The simulated probe distribution profiles were normalized based on the value present at the membrane-gel ($r = 0$ μm) at the conclusion of the 30-minute simulation run.

cm²/s was derived. Figure 5.5 displays the results of 10- and 20-minute simulated probe distribution profiles that best fit the experimental mean probe intensity profiles (also included in figure).

5.3.2   Quantitative Fluorescence Profile (QFP) Analysis of CNS Tissue Access Device (TAD) Implant Groups

Included in Figure 5.6 are fluorescent micrographs of sections collected from the upper striatum region that are representative of the totality of those captured for the control (A to C) and experimental (D to F) groups. GFAP staining (Figure 5.6A,D) revealed that often the glial scar found in the cell-transplanted group was displaced a distance away from the HFM, whereas that of controls (no cells) appeared adjacent to the membrane, with astrocyte processes sometimes projected into the macrovoid spaces of the wall. ED1 staining (Figure 5.6B,E) revealed that the region between the membrane and glial scar in the cell-transplanted group sections was densely populated by reactive microglia and macrophage cells. This is in contrast to the no-cell-transplant or control group sections, where ED1 staining was typically observed only in near proximity to the m–t interface.

The degree of cellularity, based on DAPI nuclear staining (Figure 5.6A,B,D,E), found in the membrane wall region was also significantly different between groups with a greater population of cells found residing in experimental group sections. Another difference observed between the groups related to the intensity of

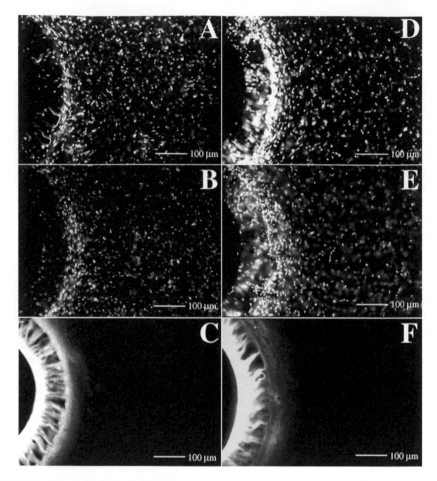

FIGURE 5.6 Representative fluorescent micrographs of upper striatum images captured from control (A to C) and experimental (D to F) group brain sections. The micrographs are composed of (A and D) DAPI (blue) and GFAP (green), (C and E) DAPI and ED1 (green), and (C and F) 70-kDa TR-dextran (red) probe distributions. **(See color insert following page 110.)**

TR-dextran probe (Figure 5.6C,F) found in the tissue region immediately adjacent to the HFM. In general, the probe distribution in experimental sections appeared more diffuse than that found in control sections.

Mean intensity profiles evaluated based on the entire set of composite QFP profiles collected for the upper striatum region appear in Figure 5.7, where results for the no-cell control (A) and cell-transplanted or experimental (B) groups are displayed in separate graphs. The 0-μm position on the x-axis in these graphs represents the m–t interface, with negative values reflecting data within the macrovoid regions of the HFM wall. When GFAP and ED1 composite profiles in data sets A and B were normalized to their respective peak intensity values, their mean profile peaks fell below the value of unity. This is indicative of variation in the scarring response between and around the circumference of sections. In contrast to the no-cells control group,

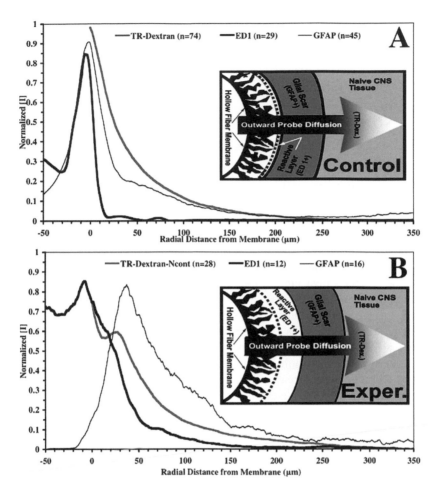

FIGURE 5.7 70-kDa TR-dextran, GFAP, and ED1 mean intensity profiles assessed from the entire data set of composite QFP profiles collected for the upper striatum region in control (A) and experimental (B) sample groups. "n" values displayed in legends represent the number of composite QFP profiles utilized in the calculation of mean intensity profiles. The TR-dextran profile for the experimental group was additionally normalized based on the proportion of probe present at the m–t interface relative to that assessed for the control group.

variations in the scarring response were especially pronounced in the cell-transplant experimental group samples (Figure 5.7B), where clearly discernable peaks are lacking in both marker profiles. In some experimental group sections the scar composition reflected that present in control samples, while in others the ED1+ reactive microglia and macrophage layer had significantly infiltrated macrovoid regions of the membrane and extended tens of microns away from the membrane. The overlap of GFAP and ED1 profiles was also indicative of radial defined spatial variations in the scarring response observed within groups. As shown in Figure 5.7, the shape of the TR-dextran profiles was clearly distinct between the groups. The profile for the experimental group was normalized based upon the proportion of probe present at the m–t interface relative to that assessed for the control. The applied normalization therefore indicates that the concentration of TR-dextran probe at the m–t interface for meningeal fibroblast implants (i.e., the experimental group) was ~75% that of the control group.

5.3.3 CNS–ECS SOLUTE DIFFUSION MODELING

5.3.3.1 Upper Striatum Mean Intensity Profile Subset Data

The insets in Figures 5.7A and 5.7B in these graphs are simplified schematics of the tissue compositional patterns utilized in the screening selection of composite QFP for the two groups. As shown in the inset in Figure 5.7A, data used for inclusion in the control group subset exhibited a host CNS response with a thin reactive microglia and macrophage layer (ED1 profile) adjacent to a glial scar layer (GFAP profile), with a regional tissue transition occurring near the m–t interface (~5 μm). Most of the control profiles examined met this description, so the subset compiled simply represents a near equal sampling from each of the five animals comprising this group. Composite QFP data used for generating the experimental subset profiles was based upon a host CNS response that presented a broad ED1$^+$ reactive microglia and macrophage layer (see inset in Figure 5.7B). Specifically, ED1 profiles that extended out from the membrane 20 to 35 μm and GFAP profiles with peak intensities located at ~40 μm from the membrane were used for the experimental group subset. The mean intensity profiles for the TR-dextran probe were generated solely based on those that coincided with QFPs included in subsets.

5.3.3.2 Evaluation of *In Vivo* Hollow Fiber Membrane (HFM) Diffusivity

Figure 5.8 displays the normalized mean intensity profile results for the 70-kDa TR-dextran probe across the membrane wall region. The x-axis has been defined such that the inner surface of the membrane is equal to 0 μm. Examination of the profile suggests that at the end of 30-minute *in vivo* diffusion runs, the concentration of probe at the outer surface of the membrane was ~10% of that present at the intralumenal surface. An effective membrane diffusivity value of $D_{memb} = 0.9 \times 10^{-8}$ cm^2/s was estimated when this relationship was used as the primary determinant for curve fitting. Probe distribution profile results from simulations performed using this value also appear in Figure 5.8. For comparison, Figure 5.8 also contains distributions representative of simulations that employed the membrane diffusion coefficients for the naïve membrane (6.0 × 10^{-8} cm^2/s) and pure water (3.8 × 10^{-7} cm^2/s).

5.3.3.3 Simulation Analysis of Control and Experimental Group 30-Minute Probe Distributions

Figures 5.9 and 5.10 display the simulated probe distribution fits (A) and the tissue tortuosity profiles (B) for the control and experimental groups, respectively. For comparative purposes the subset mean intensity profiles have been included in the charts. A second simulated probe distribution profile, referred to as the "black box fit" is also presented for the experimental group results. This fit was generated using

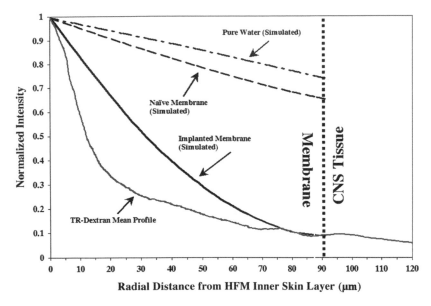

FIGURE 5.8 70-kDa TR-dextran mean intensity profile ($n = 29$) across the transmembrane region of the PAN-PVC hollow fiber membrane sections collected from the upper striatum region of rat brain. Simulated 30-minute 70-kDa dextran solute distribution profiles are also displayed for simulations run using apparent membrane diffusivity values based on probe diffusion in pure water, diffusivity characteristics from previous evaluation of naïve membrane, and an apparent membrane diffusion coefficient that resulted in a close fit to the extralumenal:intralumenal ratio of probe observed in experimental samples. The simulated probe distribution profiles were normalized based on the value at the intralumenal surface of the membrane ($r = 0$ µm) at the conclusion of the 30-minute simulation runs.

an average apparent diffusion coefficient to delineate solute transport through the affected tissue region (i.e., reactive microglia and macrophage and astrocyte layers). It should be noted that all simulated data profiles presented here resulted from simulation runs where the probe concentration in the intralumenal chamber were assumed to be constant (i.e., infinite reservoir model).

The "best-fit" TR-dextran profiles for the extents of reactive microglia and macrophage layers were obtained for 2 µm and 28 µm for the control and experimental groups, respectively. Recalling that decreases in tortuosity represent increases in tissue diffusivity (see Equation 5.3), examinations of the tissue tortuosity values observed in the reactive microglia and macrophage layer for both study groups ($\lambda = 1.76$ versus 2.23 for naïve tissue) suggest this region is less resistant to 70-kDa dextran diffusion by a factor of 1.6 ($D_{micro}^{a} = 1.2 \times 10^{-7}$ cm²/s than $D_{norm}^{a} = 7.5 \times 10^{-8}$ cm²/s). Unlike the significant change in tortuosity identified in this region, the void fraction values used to produce these fits did not differ from that of naïve tissue with $\alpha_{micro} = \alpha_{norm} = 0.21$.

The radial extents of the glial scar layers delineated in best-fit simulation runs were 50 µm and 80 µm for the control and experimental groups, respectively. The values of D_{scar}^{a} for the control group ranged from 3.00×10^{-8} to 7.32×10^{-8} cm²/s, the equivalent of λ ranged from ~3.52 to ~2.25, and were specified by linearly increasing the apparent diffusivity by 0.18×10^{-8} cm²/s per node. The values of D_{scar}^{a} for

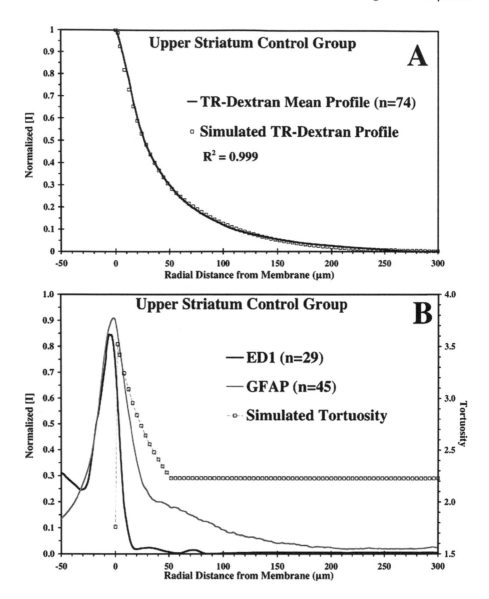

FIGURE 5.9 Results attained from finite element CNS–ECS solute diffusion simulations. (A) 70-kDa TR-dextran normalized mean intensity profile for the upper striatum control group subset along with a best fit 30-minute simulated probe distribution profile attained from simulation analysis. The simulated probe distribution profile was normalized based on the value at the m–t interface ($r = 0$ μm) at the conclusion of the 30-minute simulation run. (B) GFAP and ED1 mean intensity profiles for the control group subset along with CNS–ECS tortuosity profiles evaluated based on apparent diffusion coefficients utilized in the simulation run that produced the profile that best fit the observed dextran probe distribution. "n" values displayed in legends represent the number of composite QFP profiles utilized in the calculation of mean intensity profiles.

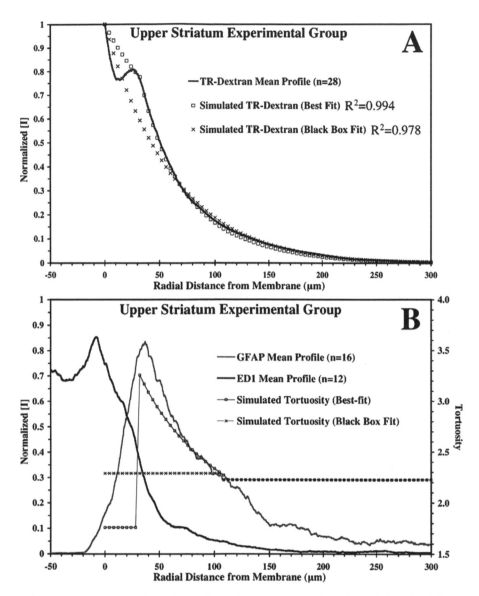

FIGURE 5.10 Results attained from finite element CNS–ECS solute diffusion simulations. (A) 70-kDa TR-dextran normalized mean intensity profile for the upper striatum experimental group subset along with 30-minute simulated probe distribution profiles for best and black box data fits. Simulated probe distribution profiles were normalized based on the value present at the m–t interface ($r = 0$ μm) at the conclusion of 30-minute simulation runs. (B) GFAP and ED1 mean intensity profiles for the experimental group subset along with CNS–ECS tortuosity profiles evaluated based on apparent diffusion coefficients utilized in simulations to produce best and black box fits for the observed dextran probe distribution. "n" values displayed in legends represent the number of composite QFP profiles utilized in the calculation of mean intensity profiles.

the experimental group were specified in a similarly linearly increasing manner and ranged from $3.50 \times 10-8$ to 7.21×10^{-8} cm^2/s, the equivalent of l ranging from ~3.26 to ~2.27, varied by ~0.093×10^{-8} cm^2/s per node. These results suggest that in control and experimental groups the leading border of the glial scar is more resistant to 70-kDa dextran diffusion by factors of 2.5 and 2.14, respectively. For both study groups a void fraction value of $\alpha_{scar} = 0.17$ was used to produce best-fit probe distribution profiles.

The results of best-fit simulations attained for both study groups also indicate an apparent diffusivity value for the unaffected brain tissue region of $D_{norm}^a = 7.5 \times 10^{-8}$ cm^2/s and a void fraction value of $\alpha_{norm} = 0.21$. These values are equal to those initially utilized for simulation analyses selected based on previously cited studies. For this reason a series of seven simulations for both study groups were performed where the value of D_{norm}^a ranged from 7.2×10^{-8} to 7.8×10^{-8} cm^2/s. In a similar manner a series of seven simulations were performed in which the value of α_{norm} ranged from 0.19 to 0.23. None of the results attained from these sets of simulations produced an improved fit to the experimentally derived TR-dextran mean intensity profiles (data not shown).

The black box fit results shown for the experimental group (Figure 5.10) treats the affected tissue region as a single entity ("micro + scar") with tortuosity and void fraction values considered to be constant. The simulated probe distribution profile shown for this fit was attained using a radial extent of 108 μm with $D_{micro+scar}^a$ 7.08×10^{-8} cm^2/s, the equivalent of $\lambda \approx 2.29$, and $\alpha_{micro+scar} = 0.21$. Based upon this result, the affected tissue region for the experimental group, overall, is slightly more resistive to the diffusion of 70-kDa dextran than unaffected tissue by a factor of 1.06.

5.4 DISCUSSION

Previous investigations regarding CNS chronically implanted (>3 weeks) therapeutic devices suggest that a variety of implants extend and may prevent the injury resolution process. This is evidenced by a sustained reactive layer consisting of activated microglia and macrophages that typically is found at the implant surfaces surrounded by hypertrophic astrocytes [3,40,50–52]. Ongoing investigations are being conducted to better understand and eventually subvert the mechanisms that result in the persistence of these reactive cells and glial scarring; however, to the best of our knowledge, it is unknown whether the solute diffusivity properties of the CNS–ECS in the reactive layer is altered by such interventions. In this chapter we describe a newly developed characterization technique capable of estimating, and thus determining, how diffusivity is affected by the cells surrounding an implanted device [38–40].

Cell encapsulation devices embedded in a 1.5% w/v agarose–gelatin gel were utilized to run *in vitro* Alexa 488–labeled streptavidin diffusivity characterization experiments. There is a long history behind the use of agar as a medium for performing such solute diffusion experiments since convective movement of a soluble probe is easily avoided [53]. Agarose, a polysaccharide derived from agar and commonly employed for gel electrophoresis, is an ideal medium because even at very low concentrations it is able to provide a relatively firm substrate at room temperature. However, employing the technique requires probe immobilization at the conclusion of test runs. Since agarose is not a proteinaceous substrate, it was necessary to incorporate gelatin into the gel mixture to enable conjugation of the probe to a structural component of the medium. For a similar reason the Alexa-labeled streptavidin was used in lieu of the lysine fixable dextran to maximize the probe-medium immobilization potential of the system.

The use of *in vitro* gel diffusion characterization experiments to validate *in vivo* investigation is well established in this area of research. Levin et al. [26] utilized radioactivity measurements of serial agar sections to validate the diffusion coefficients of solutes in brain tissue from the concentrations of probe evaluated at the end of diffusion experiments. Blasberg et al. used the diffusion of radiolabeled probes in agar gel medium to estimate free medium (i.e., pure water) diffusion coefficients for calculation of ECS tortuosity values [27]. Nicholson and Phillips [29] used the TMA$^+$-microelectrode method in agar gels to validate the technique and models employed to investigate solute diffusion in the CNS–ECS. Agarose diffusion experiments have also been employed to validate the IOI experimental method used to investigate 70-kDa dextran [32] and albumin [54] diffusivity in the CNS–ECS. Last, the Nicholson group employed agar gel medium to validate their quantitative dual-probe microdialysis technique [36].

The gel studies confirmed a number of critical experimental assumptions. First, it was necessary to establish that HFM-based delivery of solute probe to an adjacent medium was attributable to diffusive- and not convective-driven transport processes. Second, we validated that fluorescent microscopy imaging of sample sections provided an adequate foundation for extracting solute probe distribution profiles for QFP analysis. Third, light-field correction and background subtraction processing of raw-intensity pixel images for determining relative fluorescent intensity had to be assessed. Fourth, a linear correlation between relative fluorescence intensity and solute probe concentration had to be established so that QFP analysis could be employed for assessing radial probe distributions from captured images. Last, we had to demonstrate that the finite element model employed here would reliably interpret solute diffusivity properties based on replication of extracted sample probe distribution profiles.

It is our belief that the results presented here for the gel diffusivity study validate the robustness of the methodology employed in previously described studies [38]. The rationale for this conclusion stems from examination of the quality of the curve fits between actual and simulated 10- and 20-minute probe distribution profiles and the estimated effective diffusion coefficient yielded through the methods applied in the example cited here (see Figure 5.5). Recall that the theoretical equation for protein diffusion yields a value of $D^w = 7.54 \times 10^{-7}$ cm^2/s as a rough approximation of the diffusivity of streptavidin in free medium. Based on the results of our example, we estimated an effective free diffusion coefficient of $D_{eff} = 6.78 \times 10^{-7}$ cm^2/s for streptavidin in our gel system. This would seem to be reasonable since our value only differs from the theoretical approximation by ~10%. The fact that both 10- and 20-minute simulations produce the same estimated value implies the efficacy of the solute diffusion finite element model for assessing diffusivity based on a transient-state probe distribution profile.

The gel model was also capable of producing profiles reflective of diffusion in the actual system, a surprising observation given the relative simplicity of the gel medium system being simulated. The solute probe intensity profile results obtained from the actual gel diffusivity experiments suggest that the QFP analysis technique yielded data that accurately reflected the relative quantities of probe present in samples. In addition, the quality of simulated fits further supports the assumption that a linear correlation exists between relative concentrations of solute probe and captured fluorescent probe intensity, at least for the range of probe quantities used in these experiments.

Labeled dextrans have been widely utilized to examine properties related to the ECS and transport characteristics of CNS tissues [35,54–60]. The popularity of dextran solutes for probing the ECS of CNS tissues relates primarily to a number of *in vivo* properties that are ideal for this application. It is important that a solute diffusion probe not undergo specific cellular uptake or permeate cell membranes since such phenomena add additional degrees of complexity to the interpretation of data. The totality of studies to date suggest that dextran diffusion is restricted to the extracellular pathways of CNS tissue and does not undergo specific uptake by cells in this region, at least over short incubation periods [33]. In addition, the 70-kDa dextran probe used in this study has been shown to be both stable [55,56,61] and nontoxic [55,56,62] even when brain tissue is directly exposed to concentrations as high as 65 mg/ml [58], which is 260 times as concentrated as the probe solution we used to perform our *in vivo* tissue diffusivity investigations [38]. Previous investigations have also shown that dextrans do not bind with proteins *in vivo* [61]. This is important since such associations would alter the diffusive mobility of the solute probe.

In addition to specific cellular uptake, it is important to consider other routes of tissue diffusion probe elimination that may be present in the system. In this regard it has been demonstrated that dextran, regardless of molecular weight, is unable to penetrate the blood–brain barrier [56]. From direct injection studies it has also been established that the 70-kDa dextran solute has a very slow elimination rate once present in the ECS regions of the brain [57]. By modeling the elimination kinetics of the injected probe as a first-order process, an estimated half-life of $t_{1/2} \approx 90$ h for 70-kDa dextran in brain tissue was determined. For a first-order process this half-life value is the equivalent of an elimination rate constant k_e of $\approx 2.139 \times 10^{-8}$ s^{-1}. When this value for k_e is inserted into the first-order rate equation [$C/C_0 = \exp(-k_e t)$] along with t = 30 min, representative of potential exposure times used in our method, a retention value of ~ 0.996 is obtained. This rough estimate suggests that the percentage of dextran probe that is potentially eliminated during the course of an experiment of short duration is <0.4% since the solute present at the end of the 30-minute runs had resided in the tissue region less than this period.

The validity of the tissue diffusivity results obtained using this method are mainly dependent on the solute distribution profiles that were measured and assumed as accurately representative of the true distributions present at the conclusion of diffusion runs. It was previously reported that aldehyde fixation alone is capable of adequately immobilizing 70-kDa dextran probes in the ECSs of brain tissue [55,58] as well as non-CNS tissues [62]. The decision to utilize lysine-fixable 70-kDa TR-dextrans was, in part, made to further ensure prevention of postmortem diffusion of the dextran solute probe. Freezing the mobility of the solute is especially critical for this investigation, given the level of spatial resolution required to distinguish minute changes in the diffusivity characteristics across a thin band of tissue. A second reason for selecting a lysine-fixable probe was to enable evaluation of relative probe concentrations in the membrane region. By using a lysine-fixable dextran, immobilization on PAN-PVC membrane surfaces is accommodated through conjugation with adsorbed protein on the membrane surface.

In the current method, a high degree of consistency is maintained in the measurement of tissue solute concentration distributions. Among the ways we addressed this issue was through implementing a strict set of guidelines regarding the capturing of images, such as constant capture times, utilization of the blue spectrum (DAPI) for alignment of sections on the microscope, and by limiting exposures to avoid bleaching phenomenon. In addition, new light-field calibration images were acquired at the onset of each microscopy session. The QFP image analysis software program used to evaluate solute distribution profiles was also specifically developed for the current study application. The software facilitates the referencing of origins of individual radial sampling arrays at the m–t interface, which is the most readily distinguishable feature common in all sections analyzed. The circumferential sampling density of 1.7 degrees/profile effectively provides complete canvassing of the analysis region, given that higher profiling densities produce insignificant changes in composite point intensity standard deviation values.

Our decision to apply diffusivity model analyses to the upper striatum region is primarily based on the reasonability of the assumption that probe diffusion in the direction normal to the plane of that studied could be neglected. Additionally, a relatively large number of sections can be collected for this region since it spans a distance of nearly 1800 μm along the axial length of the HFM. Similar types of data could be taken for the cortex, and it is possible to create curvilinear HFM trajectories to probe other regions of the CNS [63]. Therefore 50-μm sectioning enables collection of an ample number of sections for applying desired immunohistochemical stains and additional sections required for primary stain controls and determination of probe distributions across the region of interest.

The degrees of cell compositional variability observed among brain tissue sections, especially for the cell-transplanted group, was anticipated based upon other reported findings regarding long-term resolution responses to implanted devices and materials [3,12,13,38–40,50–52,64,65]. The 70-kDa TR-dextran profile attained for the experimental group subset contains a valley in the EDI$^+$ reactive microglia and macrophage region that can not be the product of simple solute diffusion and may represent uptake of the solute by macrophages. It is likely that only a pseudo-matrix is present in this region since reactive microglia and macrophage cells are normally responsible for the removal of cellular debris, a process that requires these cells to break down existing components of the extracellular matrix to facilitate their migration through affected tissues [44,66,67]. Even though aldehyde fixation was applied, it is possible that tissue in this region was loosely attached, perhaps

contributing to the observed frequencies of tissue detachment from the m–t interface. Under this circumstance, during sectioning of brain samples or staining of
sections, indiscernible, yet significant, fragments of the tissue may have detached
along with bound dextran probe. If this phenomenon did occur, then the intensity
of the ED1 profile would also be depressed in this region. The approach taken to
model this region was based on the assumption that the TR-dextran probe intensities found at the m–t interface and the peak that lies adjacent to the valley (~40 μm)
represented reliable values for the concentrations of probe present in these locations
at the end of 30-minute diffusion runs. Based upon this two-point treatment of the
data, it was only appropriate to apply a constant apparent diffusivity value for this
region in simulation analyses.

A similar two-point treatment was applied in the determination of solute probe
effective diffusivity through the membrane where intensity values present at the
intralumenal surface and m–t interface were used to assess the fit of simulated data.
This is clearly apparent in Figure 5.9 since it is at these radial positions (0 and 90
μm) that the simulated curve intersects the actual TR-dextran mean intensity profile
evaluated from QFP analyses. The results of simulations based on this approach
suggest that the diffusivity of the membrane towards the 70-kDa dextran decreased
to approximately 15% of its naïve (i.e., preimplantation) diffusivity (D_{memb} = 0.9 ×
10^{-8} cm^2/s versus 6.0 × 10^{-8} cm^2/s). This significant decline observed in the permeability of the membrane is likely primarily the result of protein fouling of the pores
comprising the permselective skin layer and perhaps to a small degree owing to
cellular infiltration of the macrovoid regions of the membrane.

Several aspects of the tissue diffusivity results lend credibility to the techniques
that were employed. As depicted in the results presented in Figures 5.9A and 5.10A,
the overall similarity of actual and simulated probe distribution profiles attained
for both study groups suggest the robustness of the finite element model employed
in this study. Additionally, the apparent diffusivity values derived from simulation
analyses for the unaffected CNS tissue regions ($r \geq 52$ μm and $r \geq 108$ μm for the
control and experimental groups, respectively) were D_{norm}^a = 7.5 × 10^{-8} cm^2/s for
both study groups. Apparent diffusivity values attained for the reactive microglia
and macrophage and glial scar regions are also in agreement regarding the trends
observed in the results for these regions. The results for this region were in agreement with previously published findings [54], and the similarity of the findings for
the affected tissue regions together suggest the reliability of the tissue diffusivity
characterization methodology employed in the current study.

Tortuosity (λ), as related to CNS–ECS solute diffusion, is generally considered by investigators as composed of "geometric" and "viscosity" components [68].
While the geometric component reflects the increased distance a solute travels as it
circumnavigates nonlinear extracellular pathways, the viscosity component originates from frictional forces imposed on solutes by the presence of large matrix proteins and other cell-membrane-associated moieties. IOI-based studies employing
dextran solute diffusion probes have also revealed that ECS tortuosity is dependent
on solute molecular weight, with λ values increasing from 1.77 to 2.25 as the MW
increases from 3 to 70 kDa [32]. If the geometric component remains relatively
consistent for each of solute MWs characterized, a reasonable assumption since the

tissue cellular compositions and morphologies did not vary between probes, then the results of the IOI study may suggest that the viscosity component is the primary determinant of overall tortuosity for larger MW solutes.

Characterization of reactive gliosis from the perspective of macromolecule diffusivity properties has only recently been examined [38]. This study and the results presented here suggest that the permeability of a glial scar is most restrictive at its leading edge (λ values exceeding 150% and 140% of normal) and gradually increases in the transition to unaffected tissue. The presence of a denser extracellular matrix architecture in this region, which would increase the viscosity component of tortuosity, is a characteristic that has previously been associated with astrogliosis [30,69]. We also observed a decreased α value (0.17 versus 0.21) in the glial scar region for both groups. This may be the result of the morphology of cells present in the glial scar region, which is generally described by densely packed reactive astrocytes with hypertrophied and tightly interdigitating processes [43]. Important to the development of CNS implantation devices is the elucidation of the significance of a persistent reactive microglia and macrophage region on the diffusivity properties. In this respect it is clear, based on our results, that the extent of the glial scar, distinguished solely on the basis of diffusivity, was larger when this persistent reactive layer was present (80 versus 50 μm).

To our knowledge, the diffusivity properties of ED1⁺-reactive microglia and macrophage tissue are completely unknown. Our results suggest that diffusivity of the 70-kDa dextran probe in this region was significantly higher with an estimated tortuosity of ~79% of normal (D_{micro}^{a} = 1.2 × 10⁻⁷ cm²/s versus D_{norm}^{a} = 7.5 × 10⁻⁸ cm²/s). Surprisingly, the value of the void volume fraction observed in this region was similar to that of normal tissue (α = 0.21). This finding suggests a decrease in the viscosity component of tortuosity that may indicate the presence of a less dense, loosely integrated extracellular matrix architecture in this region. It is also potentially attributable to decreases in the geometric component of tortuosity since the morphology of microglia and macrophage cells, and therefore extracellular spacial characteristics, vary from that of normal CNS tissue.

The methodology described here is unique compared to methods previously used to examine tissue diffusivity properties. The TMA⁺-microelectrode and dual-probe microdialysis techniques rely upon measurements of local point concentrations versus time and are therefore inherently limited to the evaluation of average apparent diffusivities across the bulk region over which the probes span. The IOI and multiphoton microscopy techniques assess changes in probe distributions over time. The resolution of probe profiles in these methods was sacrificed in lieu of evaluating a time series of dynamic distributions from which only an average apparent diffusivity of a region is estimated. While similar to our technique in that the distribution of probe occurs at a single time point, efforts to employ the radio-label–autoradiography techniques to attain profiles of similar resolution to those achieved in this study have yet to be reported. By using QFP analysis in conjunction with finite element simulations, we were able to produce a tissue diffusivity mapping for the region of interest.

Finally, it is interesting to note that the black box fit (see Figure 5.7), representative of an average apparent tissue diffusivity value evaluated for the entire affected brain tissue region, suggests transport properties comparable to unaffected tissue regions with $D^a_{micro+scar} = 7.08 \times 10^{-8}$ cm^2/s versus $D^a_{norm} = 7.5 \times 10^{-8}$ cm^2/s (in terms of tortuosity $\lambda \approx 2.29$ versus 2.23). Implicit in this result is that care should be taken in the interpretation of results observed from diffusivity measurements derived from techniques based on the two-point methodology. The current study revealed opposite directions in tissue diffusivity properties of the compositionally distinct, but adjacent, layers of tissue across a relatively small region spanning only ~100 μm.

5.5 CONCLUSION

The results of both *in vitro* gel diffusivity and *in vivo* brain tissue diffusivity studies suggest the validity of the methodology employed to examine solute diffusion properties in the ECS of CNS tissues. Solute probe distributions present in tissue and gel sections could be extracted through quantitative fluorescence profiling analysis. The quality of the simulated fits attained in comparison to extracted probe distribution profiles also supports the utility of the finite element model employed to derive tissue diffusivity properties. Based on the methodology employed, for a 70-kDa dextran solute the diffusivity of the reactive microglia and macrophage layer is significantly higher than normal tissue, while that of the glial scar region is significantly lower.

To date, a number of clinical studies have discussed the advantages of autologous fibroblasts for site-specific and sustained delivery, which can be easily collected from the affected patient's own skin over other types of cells in terms of delivering therapeutic molecules into target CNS tissue. These results suggest that even though the amounts of therapeutic molecules may be reduced by an enhanced foreign body response to fibroblasts implanted in brain tissue, therapeutic molecules below 70 kDa can still diffuse into adjacent target CNS tissue. With the advances in genetic engineering, we may be able to suppress the release of naturally produced profibrogenic cytokines, thus reducing the level of the foreign body response to the transplanted cells and improve the efficacy of therapeutic molecule delivery using such transplanted cell types. Alternatively, it may be wise to identify cell types that produce less of an encapsulation response for use as sustained-release vehicles in adult brain tissue.

ABBREVIATIONS

λ	Tortuosity Parameter
CNS	Central nervous system
D^w	Diffusion coefficient in water
D^a	Apparent diffusion coefficient
DAPI	Nuclear dye 4',6-diamidino-2-phenylindole, dihydrochloride
ECS	Extracellular space
ED1	Antigenic marker that discriminates macrophages from microglia
GFAP	Glial fibrillary acidic protein
HFM	Hollow fiber membrane

ISM	Ion selective membrane
PAN-PVC	Poly(acrylonitrile-vinyl chloride)
PBS	Phosphate buffered saline
QFP	Quantitative fluorescence profile
TAD	Tissue access device
TMA+	Tetramethyl ammonium
TR	Texas Red

REFERENCES

1. Emerich, D. F., Tracy, M. A., Ward, K. L., Figueiredo, M., Qian, R., Henschel, C., and Bartus, R. T. Biocompatibility of poly (DL-lactide-co-glycolide) microspheres implanted into the brain. *Cell Transplant.* 8(1), 47–58, 1999.
2. Winn, S. R., Aebischer, P., and Galletti, P. M. Brain tissue reaction to permselective polymer capsules. *J. Biomed. Mater. Res.* 23(1), 31–44, 1989.
3. Turner, J. N., Shain, W., Szarowski, D. H., Andersen, M., Martins, S., Isaacson, M., and Craighead, H. Cerebral astrocyte response to micromachined silicon implants. *Exp. Neurol.* 156(1), 33–49, 1999.
4. Yuen, T. G., Agnew, W. F., and Bullara, L. A. Tissue response to potential neuroprosthetic materials implanted subdurally. *Biomaterials* 8(2), 138–41, 1987.
5. Tresco, P. A., Winn, S. R., Tan, S., Jaeger, C. B., Greene, L. A., and Aebischer, P. Polymer-encapsulated PC12 cells: long-term survival and associated reduction in lesion-induced rotational behavior. *Cell Transplant.* 1(2-3), 255–64, 1992.
6. Kruger, S., Sievers, J., Hansen, C., Sadler, M., and Berry, M. Three morphologically distinct types of interface develop between adult host and fetal brain transplants: implications for scar formation in the adult central nervous system. *J. Comp. Neurol.* 249(1), 103–16, 1986.
7. Del Bigio, M. R. and Fedoroff, S. Short-term response of brain tissue to cerebrospinal fluid shunts *in vivo* and *in vitro*. *J. Biomed. Mater. Res.* 26(8), 979–87, 1992.
8. Sekhar, L. N., Moossy, J., and Guthkelch, A. N. Malfunctioning ventriculoperitoneal shunts. Clinical and pathological features. *J. Neurosurg.* 56(3), 411–16, 1982.
9. Crul, B. J. and Delhaas, E. M. Technical complications during long-term subarachnoid or epidural administration of morphine in terminally ill cancer patients: A review of 140 cases. *Reg. Anesth.* 16(4), 209–13, 1991.
10. Geller, H. M. and Fawcett, J. W. Building a bridge: Engineering spinal cord repair. *Exp. Neurol.* 174(2), 125–36, 2002.
11. Brown, W. J., Babb, T. L., Soper, H. V., Lieb, J. P., Ottino, C. A., and Crandall, P. H. Tissue reactions to long-term electrical stimulation of the cerebellum in monkeys. *J. Neurosurg.* 47(3), 366–79, 1977.
12. Biran, R., Martin, D. C., and Tresco, P. A. Neuronal cell loss accompanies the brain tissue response to chronically implanted silicon microelectrode arrays. *Exp. Neurol.* 195(1), 115–26, 2005.
13. Biran, R., Martin, D. C., and Tresco, P. A. The brain tissue response to implanted silicon microelectrode arrays is increased when the device is tethered to the skull. *J. Biomed. Mater. Res. A* 82A(1), 169–78, 2007.
14. Polikov, V. S., Tresco, P. A., and Reichert, W. M. Response of brain tissue to chronically implanted neural electrodes. *J. Neurosci. Methods* 148(1), 1–18, 2005.
15. Tresco, P. A., Biran, R., and Noble, M. D. Cellular transplants as sources for therapeutic agents. *Adv. Drug Deliv. Rev.* 42(1-2), 3–27, 2000.

16. Uludag, H., De Vos, P., and Tresco, P. A. Technology of mammalian cell encapsulation. *Adv. Drug Deliv. Rev.* 42(1-2), 29–64, 2000.

17. Sagen, J., Wang, H., Tresco, P. A., and Aebischer, P. Transplants of immunologically isolated xenogeneic chromaffin cells provide a long-term source of pain-reducing neuroactive substances. *J. Neurosci.* 13(6), 2415–23, 1993.

18. Lindner, M. D., Winn, S. R., Baetge, E. E., Hammang, J. P., Gentile, F. T., Doherty, E., McDermott, P. E., Frydel, B., Ullman, M. D., Schallert, T., et al. Implantation of encapsulated catecholamine and GDNF-producing cells in rats with unilateral dopamine depletions and Parkinsonian symptoms. *Exp. Neurol.* 132(1), 62–76, 1995.

19. Emerich, D. F., Bruhn, S., Chu, Y., and Kordower, J. H. Cellular delivery of CNTF but not NT-4/5 prevents degeneration of striatal neurons in a rodent model of Huntington's disease. *Cell Transplant.* 7(2), 213–25, 1998.

20. Lindner, M. D. and Emerich, D. F. Therapeutic potential of a polymer-encapsulated L-DOPA and dopamine-producing cell line in rodent and primate models of Parkinson's disease. *Cell Transplant.* 7(2), 165–74, 1998.

21. Starr, P. A., Vitek, J. L., and Bakay, R. A. Ablative surgery and deep brain stimulation for Parkinson's disease. *Neurosurgery* 43(5), 989–1013; discussion 1013–15. 1998.

22. Barolat, G. and Sharan, A. D. Future trends in spinal cord stimulation. *Neurol. Res.* 22(3), 279–84, 2000.

23. Alo, K. M. and Holsheimer, J. New trends in neuromodulation for the management of neuropathic pain. *Neurosurgery* 50(4), 690–703; discussion 703–4, 2002.

24. Hochberg, L. R., Serruya, M. D., Friehs, G. M., Mukand, J. A., Saleh, M., Caplan, A. H., Branner, A., Chen, D., Penn, R. D., and Donoghue, J. P. Neuronal ensemble control of prosthetic devices by a human with tetraplegia. *Nature* 442(7099), 164–71, 2006.

25. Fenstermacher, J. D., Li, C. L., and Levin, V. A. Extracellular space of the cerebral cortex of normothermic and hypothermic cats. *Exp. Neurol.* 27(1), 101–14, 1970.

26. Levin, V. A., Fenstermacher, J. D., and Patlak, C. S. Sucrose and insulin space measurements of cerebral cortex in four mammalian species. *Am. J. Physiol.* 219(5), 1528–33, 1970.

27. Blasberg, R. G., Patlak, C., and Fenstermacher, J. D. Intrathecal chemotherapy: brain tissue profiles after ventriculocisternal perfusion. *J. Pharmacol. Exp. Ther.* 195(1), 73–83, 1975.

28. Krewson, C. E., Klarman, M. L., and Saltzman, W. M. Distribution of nerve growth factor following direct delivery to brain interstitium. *Brain Res.* 680(1-2), 196–206, 1995.

29. Nicholson, C. and Phillips, J. M. Ion diffusion modified by tortuosity and volume fraction in the extracellular microenvironment of the rat cerebellum. *J. Physiol.* 321, 225–57, 1981.

30. Roitbak, T. and Sykova, E. Diffusion barriers evoked in the rat cortex by reactive astrogliosis. *Glia* 28(1), 40–48, 1999.

31. Sykova, E. and Chvatal, A. Glial cells and volume transmission in the CNS. *Neurochem. Int.* 36(4-5), 397–409, 2000.

32. Nicholson, C. and Tao, L. Hindered diffusion of high molecular weight compounds in brain extracellular microenvironment measured with integrative optical imaging. *Biophys. J.* 65(6), 2277–90, 1993.

33. Tao, L. Effects of osmotic stress on dextran diffusion in rat neocortex studied with integrative optical imaging. *J. Neurophysiol.* 81(5), 2501–7, 1999.

34. Prokopova-Kubinova, S., Vargova, L., Tao, L., Ulbrich, K., Subr, V., Sykova, E., and Nicholson, C. Poly[N-(2-hydroxypropyl)methacrylamide] polymers diffuse in brain extracellular space with same tortuosity as small molecules. *Biophys. J.* 80(1), 542–48, 2001.

35. Stroh, M., Zipfel, W. R., Williams, R. M., Webb, W. W., and Saltzman, W. M. Diffusion of nerve growth factor in rat striatum as determined by multiphoton microscopy. *Biophys. J.* 85(1), 581–88, 2003.

36. Chen, K. C., Hoistad, M., Kehr, J., Fuxe, K., and Nicholson, C. Quantitative dual-probe microdialysis: Mathematical model and analysis. *J. Neurochem.* 81(1), 94–107, 2002.

37. Hoistad, M., Chen, K. C., Nicholson, C., Fuxe, K., and Kehr, J. Quantitative dual-probe microdialysis: evaluation of [3H]mannitol diffusion in agar and rat striatum. *J. Neurochem.* 81(1), 80–93, 2002.

38. Kim, Y. T., Bridge, M. J., and Tresco, P. A. The influence of the foreign body response evoked by fibroblasts transplantation on soluble factor diffusion in surrounding brain tissue. *J. Control. Release* 118(3), 340–47, 2007.

39. Kim, Y. T., Hitchcock, R., Broadhead, K. W., Messina, D. J., and Tresco, P. A. A cell encapsulation device for studying soluble factor release from cells transplanted in the rat brain. *J. Control. Release* 102(1), 101–11, 2005.

40. Kim, Y.-T., Hitchcock, R. W., Bridge, M. J., and Tresco, P. A. Chronic response of adult rat brain tissue to implants anchored to the skull. *Biomaterials* 25, 2229–37, 2003.

41. Broadhead, K. W. and Tresco, P. A. Effects of fabrication conditions on the structure and function of membranes formed from poly(acrylonitrile-vinylchloride). *J. Membr. Sci.* 147(2), 235–45, 1998.

42. Bridge, M. J., Broadhead, K. W., Hlady, V., and Tresco, P. A. Ethanol treatment alters the ultrastructure and permeability of PAN-PVC hollow fiber cell encapsulation membranes. *J. Membr. Sci.* 195(1), 51–64, 2001.

43. Norton, W. T., Aquino, D. A., Hozumi, I., Chiu, F. C., and Brosnan, C. F. Quantitative aspects of reactive gliosis: A review. *Neurochem. Res.* 17(9), 877–85, 1992.

44. Flaris, N. A., Densmore, T. L., Molleston, M. C., and Hickey, W. F. Characterization of microglia and macrophages in the central nervous system of rats: definition of the differential expression of molecules using standard and novel monoclonal antibodies in normal CNS and in four models of parenchymal reaction. *Glia* 7(1), 34–40, 1993.

45. Nicholson, C. and Rice, M. E. The migration of substances in the neuronal microenvironment. *Ann. NY Acad. Sci.* 481, 55–71, 1986.

46. Nicholson, C. Structure of extracellular space and physiochemical properties of molecules governing drug movement in brain and spinal cord. In *Spinal Drug Delivery*, Yaksh, T. L., ed. Elsevier Science, Amsterdam, 1999, pp. 253–69.

47. Granath, K. A. Solution properties of branched dextrans. *J. Colloid Int. Sci.* 13, 304–10, 1958.

48. Nicholson, C. and Sykova, E. Extracellular space structure revealed by diffusion analysis. *Trends Neurosci.* 21(5), 207–15, 1998.

49. Tyn, M. T. and Gusek, T. W. Prediction of diffusion coefficients of proteins. *Biotech. Bioeng.* 35, 327–38, 1990.

50. Stensaas, S. S. and Stensaas, L. J. The reaction of the cerebral cortex to chronically implanted plastic needles. *Acta Neuropathol.* (Berlin) 35(3), 187–203, 1976.

51. Benveniste, H. and Diemer, N. H. Cellular reactions to implantation of a microdialysis tube in the rat hippocampus. *Acta Neuropathol.* (Berlin) 74(3), 234–38, 1987.

52. Szarowski, D. H., Andersen, M. D., Retterer, S., Spence, A. J., Isaacson, M., Craighead, H. G., Turner, J. N., and Shain, W. Brain responses to micro-machined silicon devices. *Brain Res.* 983(1-2), 23–35, 2003.
53. Schantz, E. J. and Lauffer, M. A. Diffusion measurements in agar gel. *Biochemistry* 1, 658–63, 1962.
54. Tao, L. and Nicholson, C. Diffusion of albumins in rat cortical slices and relevance to volume transmission. *Neuroscience* 75(3), 839–47, 1996.
55. Olsson, Y., Svensjo, E., Arfors, K. E., and Hultstrom, D. Fluorescein labelled dextrans as tracers for vascular permeability studies in the nervous system. *Acta Neuropathol.* (Berlin) 33(1), 45–50, 1975.
56. Tervo, T., Joo, F., Palkama, A., and Salminen, L. Penetration barrier to sodium fluorescein and fluorescein-labelled dextrans of various molecular sizes in brain capillaries. *Experientia* 35(2), 252–54, 1979.
57. Dang, W. and Saltzman, W. M. Dextran retention in the rat brain following release from a polymer implant. *Biotechnol. Prog.* 8(6), 527–32, 1992.
58. Ohata, K. and Marmarou, A. Clearance of brain edema and macromolecules through the cortical extracellular space. *J. Neurosurg.* 77(3), 387–96, 1992.
59. Jansson, A., Mazel, T., Andbjer, B., Rosen, L., Guidolin, D., Zoli, M., Sykova, E., Agnati, L. F., and Fuxe, K. Effects of nitric oxide inhibition on the spread of biotinylated dextran and on extracellular space parameters in the neostriatum of the male rat. *Neuroscience* 91(1), 69–80, 1999.
60. Hrabetova, S., Hrabe, J., and Nicholson, C. Dead-space microdomains hinder extracellular diffusion in rat neocortex during ischemia. *J. Neurosci.* 23(23), 8351–59, 2003.
61. Rutili, G. and Arfors, K. E. Fluorescein-labelled dextran measurement in interstitial fluid in studies of macromolecular permeability. Microvasc. Res. 12(2), 221–30, 1976.
62. Thorball, N. FITC-dextran tracers in microcirculatory and permeability studies using combined fluorescence stereo microscopy, fluorescence light microscopy and electron microscopy. *Histochemistry* 71(2), 209–33, 1981.
63. Dubach, M., Anderson, M. E., and Tresco, P. A. Extended local access fibers: adjustable treatment of deep sites in the brain. *J. Neurosci. Methods* 85(2), 187–200, 1998.
64. Collias, J. C. and Manuelidis, E. E. Histopathological changes produced by implanted electrodes in cat brains. *J. Neurosurg.* 14(3), 302–28, 1957.
65. Stensaas, S. S. and Stensaas, L. J. Histopathological evaluation of materials implanted in the cerebral cortex. *Acta Neuropathol.* (Berlin) 41(2), 145—55, 1978.
66. Banati, R. B., Gehrmann, J., Schubert, P., and Kreutzberg, G. W. Cytotoxicity of microglia, *Glia* 7(1), 111–18, 1993.
67. Norton, W. T. Cell reactions following acute brain injury: A review. *Neurochem. Res.* 24(2), 213–18, 1999.
68. Rusakov, D. A. and Kullmann, D. M. A tortuous and viscous route to understanding diffusion in the brain. *Trends Neurosci.* 21(11), 469–70, 1998.
69. Sykova, E., Vargova, L., Prokopova, S., and Simonova, Z. Glial swelling and astrogliosis produce diffusion barriers in the rat spinal cord. *Glia* 25(1), 56–70, 1999.

Part IV

6 A Molecular Perspective on Understanding and Modulating the Performance of Chronic Central Nervous System (CNS) Recording Electrodes

Wei He and Ravi V. Bellamkonda

CONTENTS

6.1   INTRODUCTION

Successfully interfacing the CNS with external electronics holds great potential in improving the quality of life for patients with sensory and motor dysfunctions. The impact is already evident in the profound clinical applications of cochlear implants and deep brain stimulations [1,2]. More recently, the use of an invasive electronic brain implant, also known as a neuromotor prosthesis, to help a patient paralyzed by a tetraplegic spinal cord injury has been reported [3]. In the study, a 96-electrode array was implanted into the patient's motor cortex to establish a brain–computer interface, where the patient could move a cursor to issue different instructions by thoughts of such motions. Clearly, it adds credibility to the enormous benefits that the interface technology could bring as a potential new therapy to restore independence for those severely disabled patients. Additionally, this interface technology may have significant implications for fundamental studies in neuroscience to understand normal physiology, pathology, or treatment of disorders such as epilepsy.

A key component in such interface technology is the electrode, which is usually placed inside the CNS tissue to record neural impulses. These "spikes" will subsequently be translated into commands for external electronic devices. Currently, several types of electrodes are being used for research purposes, including microwires [4], glass electrodes [5], polymeric electrodes [6], and silicon micromachined implants [7,8]. Among these designs, microwires and silicon electrodes are the most popular. Microwires are well-established, metal-based, tip-recording electrodes. Their features include the ability to record large numbers of single units and ease of fabrication. However, they lack precise positioning inside the tissue. In comparison, silicon micromachined electrodes allow for greater control over electrode placement *in vivo*, as well as precise and versatile electrode design to accommodate signal recordings at different depths. This, however, comes at the price of a sophisticated multistep fabrication process. There are two prominent silicon electrodes in the field, widely known as the Utah electrode array (UEA) and the Michigan probe. The UEA is a three-dimensional array of needle-like structures with recording sites located at the tips, while the Michigan probe is a thin-film planar array with recording sites spaced out along the electrode shank. Our discussion will mainly refer to the silicon electrode, as such technology has the potential to enable precise dense sampling that will permit detailed mapping of the nervous system and improve the development of prosthetic devices.

Despite the aforementioned therapeutic potentials of electrode interface technology, many obstacles still need to be overcome to make this technology a clinical reality. First, it is not clear for how long these electrodes can record neural activity *in vivo*. Ideally, the electrode should remain functionally stable in the CNS indefinitely to achieve significant improvements in the lives of disabled patients. In reality, however, the neuronal recording tends to fade over time, for unknown reasons. For example, only 4 out of 11 electrodes remain functioning 6 months after the electrode implantation in cat sensory cortex [9]. Some speculate that the observed loss in recording ability is closely correlated to the adverse tissue response, which is characterized by the formation of glial scar and electrode encapsulation. The biocompatibility of implanted electrodes has been an area of intense research to improve recording reliability [10]. Another challenge for chronic CNS electrode performance is the risk of infection introduced by wires that penetrate the skull and skin to connect the electrode to external hardware. To tackle this problem, effort has been devoted to develop on-chip circuitry and wireless fully implantable micro-systems [11].

Since interactions between the CNS tissue and the electrode are critical in determining the functional performance of the electrode, strategies are being investigated to improve electrode biocompatibility. Several factors must be considered when it comes to addressing the biocompatibility issue. First, the choice of electrode material has an important bearing on its performance and longevity. Ideally, the material should be nontoxic, stable, instigate minimal to no host response, and have the desired electrical properties. Silicon and metals are commonly used for making CNS electrodes. Although they meet the electrical criteria, their mechanical properties are likely to pose a threat to the long-term recording. As we will discuss in a later section, the large mechanical mismatch between the electrode and tissue will create a high strain field at the interface. The high strain field might in turn contribute to the sustained glial response. Therefore, studies are underway to explore alternate softer electrode substrates, such as polyimide [12]. Nevertheless, this will bring new challenges for electrode insertion, as flexible electrodes might lack the stiffness needed to penetrate the pia, resulting in buckling and potentially generating more traumatic injury to the tissue.

The second factor to consider is the size and dimensions of the electrode. Intuitively, the electrode should be made as small as possible, so that tissue damage is minimized. *In vivo* study has demonstrated that initial glial response decreases as the cross-sectional area of the electrode decreases [13]. The size problem is not insurmountable, as advanced technologies developed in the semiconductor industry could be applied in fabricating smaller electrodes. However, the size of the electrode is closely dependent on the number of recording sites. It is still a matter of debate in the field as to how many neurons should be sampled to produce effective motor outputs. Also, the electrical impedance of the electrode is inversely proportional to the surface area of the recording site, establishing a design constraint in reducing electrode size before losing electrical viability.

The third factor is the surface properties of the electrode. Like many other implants, cell interaction with the electrode is a surface phenomenon; therefore, surface properties play a key role in determining the cellular response. It is well known that cells respond to both topographical and physiochemical cues. For example, a large body of literature has shown that modifying a material surface with extracellular matrix proteins such as fibronectin, collagen, and laminin, or peptide fragment from these proteins such as RGD (Arg-Gly-Asp), IKVAV (Ile-Lys-Val-Ala-Val), or YIGSR (Tyr-Ile-Gly-Ser-Arg), can significantly improve cell adhesion, morphology, and differentiation compared to untreated surfaces. Another example is that cells behave differently on a rough surface compared to a smooth one. Surface features, such as grooves, ridges, pillars, and holes might present physical stimuli to cells in contact with the surface. The critical issue in surface modification for electrode application is to minimize interference with the intrinsic electrical properties while achieving a better cellular interaction. Two critical questions result from a review of this literature: (a) Can these findings translate to the ability to modulate tissue reaction of implantable electrodes? and (b) Can these modifications be conducted without any adverse electrical costs in electrode performance?

6.2 CELLULAR AND MOLECULAR ASPECTS OF BRAIN RESPONSE TO ELECTRODES

Inflammation is generally considered part of the reaction of vascularized living tissue to local injury. It isolates or walls off the injurious substance or process. Although the CNS is considered immune privileged, it follows a similar inflammatory response scenario against biomaterial implants, as shown in Figure 6.1. The response is generally initiated by the implant insertion and sustained by a foreign body reaction against the indwelling implant. It involves several types of cells working in concert by means of molecular mediators. Here we will give a brief overview of the main cellular and molecular players involved in the brain inflammatory response against the electrode at various stages of the process.

6.2.1 INITIAL INSERTION-INDUCED RESPONSE

6.2.1.1 Vascular Damage

The intricate vasculature network in the brain implies that there is a high probability of vascular insult accompanying electrode implantation. En route to the brain parenchyma, the electrode first encounters the meningeal layers (Figure 6.2). The meninges consist of the tough protective outer layer, the dura mater (usually removed prior to electrode insertion), the arachnoid, and the pia mater, the thin innermost layer. Arteries and veins traveling across the pial surface demand extra attention to carefully position the electrode so that large visible surface vessels can be avoided. In addition, the cortex carries the highest vascular density in the brain, up to 160 capil-

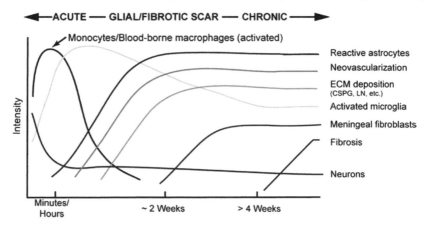

FIGURE 6.1 Hypothesized temporal sequence of inflammation and wound healing response to implanted biomaterials in the CNS. The time and intensity variables are determined by the extent of the injury, as well as the dimension, shape, surface, and bulk properties of the biomaterials.

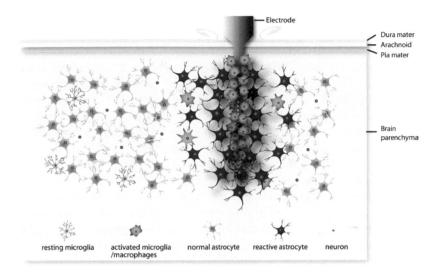

FIGURE 6.2 Cartoon showing the brain tissue response to an implanted electrode at the cellular level. Microglia activation is demonstrated by the morphological change from highly ramified to amoeboid macrophage-like. Astrocytes become reactive, as shown by the enlargement in cell size (cellular hypertrophy) and an increase in the expression of the GFAP indicated by the darker color.

laries per mm², with shorter and smaller-caliber arteries forming a compact network with the longitudinal vessels [14]. Using brain slices with fluorescently labeled vasculature, Bjornsson and coworkers [15] evaluated vascular damage during electrode insertion. Four general types of vascular damage are commonly observed: (a) fluid displacement; (b) vessel rupture; (c) vessel severing; and (d) dragging of blood vessels by the device [15]. Accompanying such vascular damage is the extravasation of serum proteins and local infiltration of cells such as neutrophils and blood-borne macrophages leading to brain edema formation [16,17].

Studies have shown that the extravasated proteins (i.e., thrombin, plasmin, and other proteases) contribute to the inflammatory response. Thrombin is a serine protease known to be an essential component of the coagulation cascade by converting fibrinogen to fibrin. Thrombin has also been suggested as a molecular trigger for astrogliosis, and a microglial activator [18–29]. Thrombin-induced cellular effects have been shown to be mediated by proteinase-activated receptors (PARs) [30,31]. In astrocytes, thrombin has been found to induce morphological changes, proliferation, secretion of endothelin-1, and release of inflammatory mediators [32–34] by activating the thrombin receptors, mainly PAR-1 [19,35,36]. PAR-1-deficient mice (PAR1^{-1-}) showed a reduced astrocytic response to cortical stab, indicating that selective activation of PAR-1 *in vivo* induces astrogliosis [28]. The molecular mechanism was proposed to be a sustained activation of the extracellular receptor kinase/mitogen-activated protein kinase (ERK/MAPK) signaling pathway [28,35]. Apart from astrocytes, thrombin also takes direct actions on microglial cells. It was demonstrated that thrombin treatment of microglia *in vitro* induces clear indications of activation including nitric oxide (NO) production, cytokine release, such as tumor necrosis factor-α (TNF-α), and increased cell proliferation [23–26], all of which are clear indications of activation. Moreover, *in vivo* studies have also shown a thrombin-induced, microglia-mediated inflammatory factor in the CNS [37]. Thrombin receptor PAR-1 contributes primarily to promoting microglial proliferation, while PAR-4 is fully responsible for activating microglia and inducing cytokine production. The signaling pathway of PAR-4 is featured by MAPK activation and subsequent nuclear factor κB (NF-κB) transcription factor activation [22–26]. In summary, plasma enzymes such as thrombin may play an important role in the CNS response to injury. It follows that a better understanding of the molecular mechanisms behind the plasma enzyme–induced CNS inflammation could provide valuable insights for potentially effective therapeutic strategies.

Besides protein extravasation into the CNS parenchyma, early compromise of the microvasculature by the penetrating electrode also leads to the influx of a variety of hematogenous cell types, including blood-borne macrophages, neutrophil granulocytes, and T-lymphocytes. Depending on the extent of vasculature damage, the cell influx could be massive and result in severe inflammatory response. The early work of Fitch and Silver [38] has demonstrated a close correlation between the tissue distribution of the upregulation of chondroitin sulfate proteoglycans (CSPGs) and the presence of activated macrophages as well as a compromised blood–brain barrier (BBB). CSPGs are a family of inhibitory molecules present in the wound healing matrix that may stall axon regeneration in the CNS after trauma. It was also observed that animals with relatively sparse inflammatory cell infiltration but clear evidence of BBB compromise showed no detectable increases in CSPG immunoreactivity. This indicates that qualitatively higher concentrations of activated macrophages are necessary to induce increases in CSPGs, even in the presence of serum components. Clearly, minimizing the access of hematogenous cell types to the CNS will aid in modulating the inflammatory response caused by chronic electrodes.

The electrode will encounter the meningeal surface before descending to the target location. Hence, there is a likelihood that some meningeal cells will be carried into the brain [14], as shown in Figure 6.3, or migrate along the electrode shank in an effort to reform the glia limitans by interacting with local astrocytes [40]. The glia limitans is a structure consisting of layers of astrocytes, layers of meningeal cells, and a layer of basal lamina in between. It is usually located parallel to the surface of the brain and spinal cord. A normal glia limitans functions as a barrier to axonal outgrowth so that axons do not grow out of the brain or spinal cord. The formation of glia limitans between the electrode and the CNS is undesirable, as its inhibitory nature could suppress neurite outgrowth toward the electrode. Moreover, by repelling neurons from the recording sites, the long-term functionality of the electrode (recall that the recording mode is a particular focus in this chapter) would be compromised.

The contribution of the initial vascular damage to the inflammatory response to implanted electrodes cannot be overemphasized. The extent of damage could be determined by the following factors: the size and geometry of the electrode, the physical insertion parameters, and the location of the insertion site relative to pial structures. Intuitively, the smaller the electrode is, the less damage it causes. While relatively little can be done to change the location of insertion, limited by the

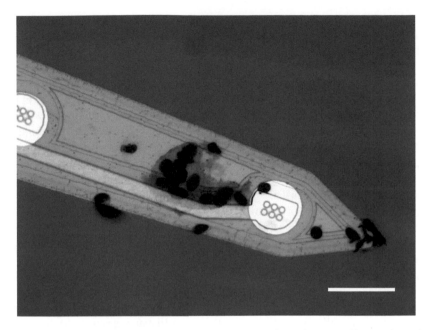

FIGURE 6.3 A vimentin+/GFAP signal was observed at the recording region of an extracted probe 4 weeks after implantation, shown here with 4,6-diamidine-2-phenylindole (DAPI) (blue) and vimentin (red) overlayed with the bright field image [39]. Scale bar = 50 μm. **(See color insert following page 110.)**

intended application, recent findings from the work of Bjornsson and coworkers [14] will help guide the selection of insertion parameters and future design of electrodes. This study evaluated the contribution of tip geometry and insertion speed to the damage caused by insertion. Within the range of speeds and tip geometries studied, faster insertion of sharp electrodes resulted in lower mean effective strain in superficial and middle regions of cortex. Insertion speed seems to play a more important role than tip geometry, with faster insertions (2000 μm s^{-1}) generally resulting in less vascular damage.

6.2.1.2 Mechanical Injury

From the cellular point of view, the initial interaction between the electrode and the brain is relatively traumatic. As the electrode traverses the brain parenchyma, it ruptures the blood vessels, disrupts the integrity of the extracellular matrix, slices the cell bodies, and displaces the tissue. However, the complete cascade of events that occur during the initial insertion, both at the cellular and molecular levels, is not yet fully understood. The consequences of initial damage are very similar to those caused by cortical stab wound. In response to the early mechanical trauma, the wound healing process is initiated. This process is characterized by the early arrival of blood-borne leukocytes, such as macrophages, via the breached vasculature and the activation of microglia, known as the resident immune cells in the nervous system (Figure 6.2). Prior to injury, microglia reside in a resting state, typically characterized as ramified with an elaborate tertiary and quaternary branch structure. The resting microglia take an active role in providing extensive and continuous surveillance of their cellular environment. As they are the first line of defense, microglia quickly become activated after CNS injury. The activation is marked by a number of characteristic events, including contraction of cellular processes and transformation from ramified into a round amoeboid macrophage-like morphology, increased proliferation, induction of immunomolecules such as major histocompatibility complex (MHC) antigen classes I and II, and changes in the pattern of cytokine and growth factor production. Although the precise factors responsible for triggering microglial activation remain to be identified, potential triggers include cellular debris generated by the injury, molecules released from dying neurons, and extravasated plasma constituents. Activated microglia will then engage in phagocytic activity, similar to the role that macrophages play in non-CNS tissue, by engulfing and digesting the debris with lytic enzymes.

The initial inflammatory response also features actions from another type of glial cell, the astrocyte (Figure 6.2). Astrocytes account for 30 to 65% of the glial cells in the CNS [41]. They play an essential role in ensuring normal neuronal activity through uptake and release of glutamate, homeostatic maintenance of extracellular ionic environment and pH, water transport, and preservation of the BBB integrity. Astrocytes are usually characterized by their star-like appearance with fine cellular processes. They respond to injury with a hallmark action, increasing the expression of the glial fibrillary acidic protein (GFAP), a fibrillary intermediate filament that is specific to astrocytes. Functionally, GFAP is essential in long-term maintenance of CNS myelination as well as stabilization of the BBB [42]. The increased expression of GFAP is marked by increases in GFAP mRNA production and is accompanied by an increase in cell volume and the caliber of proximal cell processes (cellular hypertrophy). Furthermore, astrocytes undergo dramatic biochemical and functional transformations upon activation. Collectively, these responses are referred to as reactive astrogliosis. It is generally accepted that reactive astrocytes play a key role in forming a physical barrier, commonly known as the glial scar, in an attempt to prevent injury from spreading to surrounding healthy tissue. However, in relation to its effect on electrode recording, the glial scar is considered undesirable, as it impedes regenerating neural processes coming into contact with the electrode.

6.2.1.3 Molecular Mediators

Accompanying the early cellular response is the release of a variety of molecules that participate actively in orchestrating the tissue response. Some of the molecules are secreted temporarily, while the production of others may extend into the chronic phase of the implant. Although a diverse assortment of molecular mediators is involved in the tissue response, we have chosen to focus our discussion on the expression and effects of cytokines, proteases, and reactive oxygen species.

6.2.1.3.1 Cytokines
Cytokines are low-molecular-weight glycoproteins that function as mediators of intercellular communication. Studies have shown that cytokines are rapidly secreted after CNS injuries and are essential for the initiation, propagation, and termination

of the inflammatory response [43]. Cytokines exert their actions through specific cell surface receptors in an autocrine or paracrine fashion. Because of the complexity of the cytokine network induced by injury, our discussion will focus on some of the major players, including TNF-α, interleukin-1 (IL-1), IL-6, and transforming growth factor-beta (TGF-β). The signaling pathways that are closely associated with these cytokines are the ones involving the transcription factor NF-κB and MAPKs such as p38.

6.2.1.3.1.1   TNF-α

TNF-α, a 17-kDa peptide, is one of the prototypic proinflammatory cytokines that are rapidly upregulated in the injured CNS. Activated macrophages, infiltrating through the breached vasculature, are the major cellular source for TNF-α. It has been shown to promote microglia phagocytosis as well as further production of inflammatory cytokines [44]. Activated microglia, which are the resident immune cells in the CNS, are also an early and prominent source of TNF-α. It is likely that the early expression of TNF-α is caused by the initial implantation injury. Staining against TNF-α in a stab wound lesion revealed that the immunoreactivity was predominantly located around the site of the lesion, corresponding to the location of activated microglia and macrophages [45]. Using *in situ* hybridization, we examined the spatial and temporal profiles of TNF-α expression in rats subjected to Michigan Si electrode implantation. As shown in Figure 6.4, an elevated TNF-α mRNA expression was observed right at the electrode–tissue interface 1 week after implantation, echoing staining results previously reported [45]. However, very little expression was found at the interface 4 weeks postimplantation. Previous studies have demonstrated that the transient expression of proinflammatory cytokine TNF-α is an indication of its key role in controlling the acute inflammatory response and its function in triggering secondary events [46].

TNF-α serves multiple roles in CNS response to injury. It can activate microglial cells via the autocrine loop to maintain their activated status. TNF-α also induces the proliferation of astrocytes, one of the key features of astrogliosis. In addition, studies have demonstrated that TNF-α can be directly cytotoxic to oligodendrocytes [47] (the myelin-forming cells in the CNS) and neurons [48]. Despite the deleterious role TNF-α plays in CNS tissue response to injury most of the time, studies have shown that under certain situations, TNF-α also plays a beneficial role. It appears that low levels of TNF-α may be neuroprotective [49], and such an effect could be applied indirectly via induction of growth factors such as nerve growth factor (NGF) [50].

6.2.1.3.1.2   IL-1 and IL-6

The interleukin family is composed of a growing list of members, including IL-1, IL-2, IL-3, IL-4, IL-6, IL-8, IL-10, and more. Each of them fulfills a different functional role in the immune response. In the CNS, two widely studied interleukins are IL-1 and IL-6. IL-1 exists in two forms, IL-1α and IL-1β, with the former being mostly membrane associated, while the latter is usually secreted [43]. In normal brain tissue, IL-1 is constitutively expressed at low or undetectable concentrations, at the mRNA and protein level [51]. In response to CNS injury such as electrode implantation, the expression of IL-1α and IL-1β is upregulated. Similar to TNF-

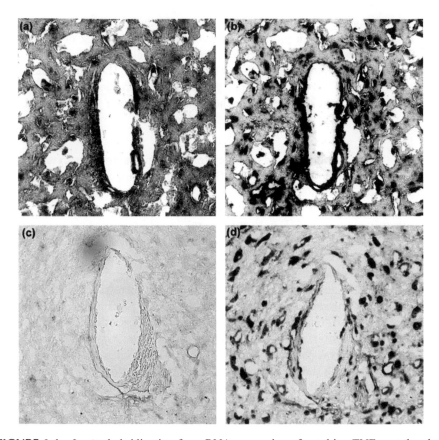

FIGURE 6.4 *In situ* hybridization for mRNA expression of cytokine TNF-α at the electrode–tissue interface. Temporally, the expression was mainly observed in the early time point, 1 week postimplantation (a), as compared to very little expression 4 weeks postimplantation (c). Spatially, the expression was highly concentrated right at the electrode–tissue interface location where activated microglia and macrophages reside, suggesting that these cells are a major source for TNF-α. Corresponding poly d(T) positive control staining (b and d) is also presented.

α, early expression of IL-1 occurs in cells of monocyte and macrophage lineage, including microglia. Studies have shown that IL-1β is rapidly produced by activated microglia within 15 min following a cortical injury [52]. IL-1 has numerous effects on glial cells and neurons. Of all the glial cells, the effect of IL-1 seems to be strongest on astrocytes. Compelling evidence links IL-1β to the induction and modulation of the astrogliosis process [53,54], featured by a repertoire of astrocyte proliferation, upregulation of GFAP expression, and increase in production of cytokines, growth factors, matrix proteins, and so on. Microglia also express receptors for IL-1β; therefore, via an autocrine feedback loop, the inflammation signal is amplified and further microglial activation is stimulated.

Like TNF-α, the effects of IL-1 in the CNS also come with a positive spin, with several recent studies highlighting the role of IL-1 in the regenerative process in the CNS. It can exert beneficial effects when released in modest concentrations [55].

It is suggested that IL-1 can improve neuronal survival and repair by directly or indirectly inducing a subset of genes primarily associated with neuronal and glial growth and survival, including IL-6, ciliary neurotrophic factor (CNTF), and NGF [56].

Unlike TNF-α and IL-1, which are mostly known for their proinflammatory properties, IL-6 has been observed to act in a proinflammatory and antiinflammatory manner. In the CNS, IL-6 is produced by microglia, astrocytes, and endothelial cells. The production is considered a downstream sequence related to TNF-α or IL-1 stimulation on cells. As IL-6 shares the signaling receptor with several growth factors (e.g., CNTF), studies have shown that IL-6 can promote neuronal survival and neurite growth [57], as well as downregulate the expression of TNF-α [58]. Together, these features are indications of the beneficial roles of IL-6 in repair and modulation of inflammation in the CNS. However, IL-6 also has proinflammatory potential. The cellular action of IL-6 is primarily on reactive astrocytes by promoting cell proliferation [59] and is believed to be involved in astrogliosis.

6.2.1.3.1.3 TGF-β

TGF-β is a multifunctional cytokine with wide-ranging effects on cell proliferation, differentiation, migration, angiogenesis, and extracellular matrix remodeling [56]. In normal CNS, TGF-β is present at low levels, but the expression is upregulated upon injury. Cellular staining for TGF-β reveals that the expression is predominantly around the site of injury and mainly colocalized with GFAP positive reactive astrocytes, indicating that astrocytes are the key cellular source of TGF-β [45]. Additionally, TGF-β can be produced by microglia as a downstream product in response to activation by the IL and TNF families [60]. As is the case with the other cytokines discussed in this section, TGF-β is considered both pro- and antiinflammatory. As a potent antiinflammatory cytokine, TGF-β has immunosuppressive effects on glial cells by inhibiting expression of proinflammatory cytokines such as TNF-α and IL-1, as well as by inhibiting glial cell proliferation [61]. However, TGF-β can also act as a proinflammatory cytokine, especially when expressed at a high concentration, to exacerbate the reactive astrogliosis and scar formation. Studies have demonstrated that, when function-blocking antibodies to TGF-β are administered, the deposition of fibrous scar tissue and the formation of a limiting glial membrane that border the lesion are significantly attenuated [62].

6.2.1.3.2 Proteases

Activated glial cells in the CNS also express proteases that exert both beneficial and harmful effects. These proteases include cathepsins, calpains, plasminogen activators, and matrix metalloproteinases (MMPs). Their physiological roles include orchestrating cell migration as well as extracellular matrix maintenance and remodeling. Here we will take a closer look at the role of MMPs in the CNS inflammatory response. The MMP family consists of over 18 members including gelatinases (MMP-2 and -9), collagenases (MMP-1, -8, -13, -18), and stromelysins (MMP-3, -7, -10, -11) [63]. All cell types in the CNS are potential sources for MMPs. A wealth of data has linked MMPs to CNS injury and inflammation. In cortical stab injury, upregulation of mRNA expression of MMPs is observed within 24 hours. Such elevation has been mainly credited to activated microglia and macrophages and reactive astrocytes. It is likely that MMP elevations postinjury are mediated by MAPK pathways or oxidative stress [64,65]. An overproduction of MMPs could be deleterious, as it contributes to BBB breakdown and therefore results in infiltration of circulating immune cells that will further amplify the inflammatory response. In addition, MMP may also degrade the parenchymal extracellular matrix protein laminin that will disrupt the laminin–neuronal interactions and contribute to neuronal death. Paradoxically, there may be some positive effects. MMPs might play a role in angiogenesis, which is essential in repair. Remodeling of the ECM by MMPs could also facilitate migration of cells to the lesion area to clean up the debris or even assist in axonal elongation.

6.2.1.3.3   Reactive Oxygen Species

Reactive oxygen species (ROS) are a class of oxygen free radicals and related molecules that are capable of exerting oxidative stress on cells when produced in excess. Nitric oxide (NO), a simple yet highly versatile molecule, is a type of ROS frequently studied and involved in a wide range of physiological as well as pathophysiological mechanisms. In normal brain tissue, production of NO is usually neutralized by cellular antioxidants. However, in response to injury, the production of NO can overpower the antioxidants, leading to oxidative stress and cellular damage. Reactive astrocytes and activated microglia have the capacity to generate NO, and the production can be induced by proinflammatory cytokines such as IL-1 and TNF-α. NO is a ubiquitous second messenger [66] that diffuses freely across cell boundaries and can damage neurons by potentiating glutamate excitotoxicity [67].

6.2.2   Tissue Response to the Chronic Presence of Electrodes

To distinguish the impact of initial mechanical injury from the chronic presence of electrodes on tissue response, Biran et al. [68] compared tissue reaction to chronically implanted microelectrodes with time-matched stab wound controls created using identical microelectrodes and an implantation technique. A striking difference was noted, with microelectrode stab wounds eliciting a subtler response that weakened with time, while indwelling microelectrodes generated a glial scar that became more compact and confined to the electrode with time. These results suggest that the initial mechanical-insertion-induced injury is transient, with the persistent presence of the electrode in the tissue accounting for the observed long-term inflammation. To some extent, this finding echoes a previous study by Szarowski et al. [13], which showed that initial injury response was a function of the implant dimensions, while the sustained injury response was independent of the implant size. This suggests that the continuous presence of the implant induces the formation of a sheath composed of reactive glial cells. Such chronic inflammation could be explained from the following perspectives.

6.2.2.1   Frustrated Phagocytosis

The common mechanism of the body defending against a foreign object is delegating macrophages to interrogate the object. These cells will then either secret lytic enzymes to degrade the object or, in the case of a nondegradable object too large to be phagocytosed, they will fuse into multinucleated foreign body giant cells and continue secretion of degradative agents such as superoxides and free radicals. The latter case is a phenomenon known as "frustrated phagocytosis." It is likely that the insoluble electrode resists the attempted removal action from activated microglia and macrophages, which means the stimulant will linger and the microglial activation will remain, leading to increased secretion of inflammatory products that further exacerbate neuronal damage.

6.2.2.2 Mechanical Strain

Chronic inflammation and glial scarring could also be instigated by the micromotion surrounding the electrode. In contrast to the softness of brain tissue, electrodes are considerably hard and rigid. There is a drastic mechanical mismatch between the electrode material, usually silicon, and the brain tissue, as the Young's moduli of bulk silicon and brain are ~200 GPa [69] and 6 kPa [70], respectively. Because of this large mismatch, brain micromotion that arises from physiological sources such as cardiac rhythm and fluctuation in respiratory pressure [71], behavioral sources such as spontaneous head movements [72], and mechanical sources such as disturbances of the implanted devices [73] could be translated into mechanical stresses and strains that impose on the tissue adjacent to the electrode, leading to compression, expansion, and even tearing of the tissue.

Recently, several groups including our own have examined the magnitude of tissue micromotion relative to a stationary implant [74] and the interfacial strains induced around the implant by micromotion [75,76]. The surface micromotion in the rodent somatosensory cortex was quantified to be on the order of 10 to 30 μm owing to pressure changes during respiration and 2 to 4 μm owing to vascular pulsatility [74]. Using finite-element modeling (FEM), the interfacial strain profiles were analyzed around a single-shank Michigan type Si microelectrode (Figure 6.5). The maximum strain occurred at the tip of the electrode in the tissue, and the strain regions could extend up to 100 μm away from the implant interface and approximately 70 μm beyond the electrode tip, with strain values decreasing exponentially as a function of distance from the interface. The strain field located around the electrode track is attributed to frictional shear stresses induced by the longitudinal displacement or the "poking" action of the electrode, which also could lead to extensive compression and tearing of the tissue. The simulation results further suggest that induced strain is affected by the extent of adhesion between the electrode and tissue. In the poor adhesion condition where the friction coefficient is finite, high strains resulted (10^{-1} to 10^{-2}) and were concentrated in a narrow region around the tip. Since the recording sites are usually located in areas toward the tip, such a strain profile is obviously undesirable, as it could cause an elevated tissue response in that region. In contrast, good adhesion representing no slipping between the electrode and tissue led to lower-magnitude strain (10^{-3} to 10^{-4}) distributed more uniformly along the entire electrode track. Although the results from these simulated studies have yet to be experimentally validated, they provide insight into the events caused by the mechanical mismatch at the interface and a basis for future improvements to

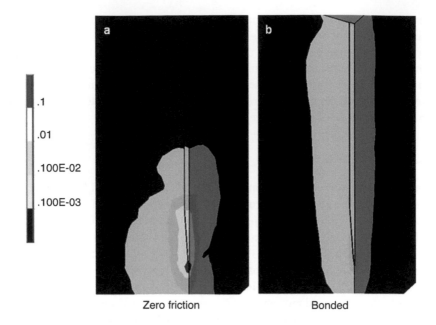

FIGURE 6.5 Von Mises strain contours (logarithmic scale) of the brain tissue on the surfaces of symmetry planes for longitudinal micromotion. (a) The zero friction coefficient case, where the significant high strain is located at the tip of the electrode. (b) The bonded case, where the peak strain is smaller and the strains are more uniformly distributed along the entire length of the electrode. **(See color insert.)**

the electrode bulk as well as surface properties to minimize mechanically induced strain that will subsequently translate into less scarring in the long term.

Thus, at the cellular level, how are these mechanical signals being transduced to cells, to what cell type(s), and with what consequences? Although there is no direct report on the effect of electrode-related mechanical strain injury on cells in the nervous system, mechanotransduction has been studied in another injury model: traumatic brain injury (TBI). Using an *in vitro* stretch model, the effects of mechanical strain injury on various brain cell types, especially astrocytes, were studied [77–84]. Astrocytes sense mechanical stress and translate it into chemical messages, such as rapid increase in intracellular free sodium and calcium, activation of phospholipases, free radical formation, and secretion of the potent astrocyte antigen endothelin, that have been implicated in the induction of gliosis. In addition, stretch-induced injury also causes delayed neuronal depolarization, which may be related to the transient neuronal dysfunction observed *in vivo* [85].

The hypothesis that continual physical insults resulting from micromotion at the electrode–brain interface could exacerbate glial scar formation, which in turn will severely compromise the recording stability of the electrode, is currently being investigated. Utilizing the *in vitro* stretch model, our group is studying the effects of sustained, low-frequency and low-magnitude strain, mimicking the actual *in vivo* situation, on astrocytes and microglia [86]. After subjecting an astrocyte culture to 24 hours of low-frequency (30 cycles/min), low-magnitude strain (5%), real-time reverse transcription-polymerase chain reaction (RT-PCR) analysis indicated that there are mild increases in mRNA expression in the following markers: TGF-β1, a cytokine that is known to enhance glial scarring and extracellular matrix deposition at high level [62]; GFAP, a characteristic indicator of astrocyte activation; and neurocan, a member of the CSPG family that is known for its role in blocking axon regeneration and is a major component of the glial scar [87]. As for microglia, although sustained strain had little effect on the gene expression of proinflammatory cytokine TNF-α, it did induce a mild upregulation in the expression of IL-1β, a very potent signaling molecule that modulates the inflammatory response. The *in vitro* results suggest that micromotion-induced mechanical strain could be perceived as a continuous stimulus and lead to glial scarring by direct stimulation of the astrocytic response or indirectly through IL-1β modulation. Future studies include investigating the mechanical strain effect in an astrocyte–microglia coculture system and exposing the culture to a longer period of strain. Additionally, it would be interesting to further elucidate the mechanical effects using organotypic slice cultures with electrodes implanted to more closely mimic the actual *in vivo* scenario.

6.2.2.3   Molecular Mediators

The chronic phase of CNS tissue response to electrodes shares a cast of molecular mediators similar to those aforementioned in the initial phase (i.e. cytokines, ROS, proteases, etc.). However, the focus of their actions might shift from engaging the defense system against the insult to restoration of tissue homeostasis and recovery and repair of the injured tissue. The cytokine balance might also tilt toward the side of antiinflammatory cytokines as opposed to initially favoring proinflammatory cytokines.

6.3   MOLECULAR STRATEGIES TO MINIMIZE THE ABOVE RESPONSES

To fully translate the enormous potential of CNS electrodes into clinical realities, one has to overcome several challenges, including the issue of long-term biocompatibility of the electrode. Clearly, a thorough understanding of the underlying biological processes behind CNS tissue response to the electrode, at the cellular and molecular level, will provide valuable implications for the development of therapeutic interventions to improve its chronic CNS performance. Although the details of such complex biological responses to CNS electrodes remain to be delineated, a consensus has been reached among the scientific community that the lack of long-term stability could be attributed to a myriad of responses that the tissue mounts against the invading electrode. As noted above, the prominent aftermath is set with the opening act of electrode insertion, which causes traumatic injury and vessel damage, followed by transition into the acute phase as the tissue attempts to repair the initial injury and remove the electrode, and eventually phased into the chronic response to the persistent inflammatory stimuli, that is, the physical presence of the indwelling electrode and micromotion-induced mechanical strain. We will briefly discuss some of the potential strategies to control CNS tissue response against the electrode. The overarching objective is to minimize the adverse inflammatory response and glial scar formation and to promote neuronal survival and outgrowth toward the electrode. The proposed strategies could potentially intervene at a molecular level.

6.3.1   STRATEGY I: BLOCKING PROTEINASE-ACTIVATED INHIBITORS (PAR) RECEPTORS

Even though the complete consequences of vascular damage by the initial electrode implantation on chronic electrode performance are not yet clear, one cannot overlook the fact that serum components that are extravasated via the compromised BBB play an indispensable role in triggering the inflammatory response. Serine proteases such as thrombin are powerful activators of glial cells in the CNS and can induce reactive gliosis *in vivo*. However, it would be unwise to completely inhibit thrombin activity, as it could trade dangerously excessive bleeding for its potential therapeutic effect. Thus, it may be beneficial to identify specific molecular targets mediating thrombin effects. A potential approach could be inactivating or blocking the receptors for thrombin expressed by the glial cells. As studies have demonstrated the involvement of the receptors PAR-1 and PAR-4 in thrombin actions on microglia and astrocytes, it is possible to employ anti-PAR1 and anti-PAR4 tactics to subdue the effects of thrombin. The available PAR-1 antagonists include peptide mimetics such as BMS-200261, RWJ-56110, RWJ-53052, and RWJ-58259 and peptidic antagonists such as RPPGF [27]. The selectivity of these antagonists has been characterized in tissues outside the CNS, but their effectiveness has not been evaluated in the CNS except for RPPGF. In comparison with PAR-1, PAR-4 is a relatively newly identified receptor for thrombin; thus, there is very limited information available for selective PAR-4 antagonists, except a synthetic compound YD-3 [88] and a pepducin type antagonist P4pal-10 [89].

Besides identifying the specific molecular targets to combat thrombin-triggered CNS inflammation to the electrode, several other issues need to be addressed as well. In particular, questions such as how to deliver the antagonists, when to release them, at what concentration, and for how long demand careful consideration. As vessel rupture takes place simultaneously with initial electrode insertion, it would be favorable to release the PAR antagonists early on. Also, as studies have suggested the biphasic impact of thrombin in the CNS (low concentrations of thrombin can be neuroprotective, while high concentrations can be deleterious), partially blocking PAR-1 and PAR-4 in the CNS may hold more promising therapeutic value than completely diminishing the PAR effects.

6.3.2   STRATEGY II: ATTENUATING INFLAMMATION

One major consequence of the inflammatory response is the formation of glial scar, which not only physically blocks neurons away from the electrode, but also electrically isolates the electrode from surrounding neurons by increased impedance. Considerable effort is being directed toward minimizing, or better, eliminating, the formation of such a barrier. The key members of the inflammation cast are microglia and astrocytes, with a supporting cast of molecular mediators such as cytokines, growth factors, and reactive oxygen species. Also, NF-κB is a pivotal transcription factor for genes that encode proinflammatory cytokines such as IL-1 and TNF-α [90]. Several antiinflammatory agents have been shown to exert powerful effects through inhibition of NF-κB activation, including the neuroimmunomodulatory peptide α-melanocyte-stimulating hormone (α-MSH) and synthetic glucocorticoids such as methylprednisolone and dexamethasone. Spataro et al. [91] found that peripheral injections of dexamethasone beginning at the day of electrode insertion and continuing for 6 days profoundly affected the early (1 week after insertion) and sustained (6 weeks after insertion) reactive responses. However, the success of systemic delivery of glucocorticoids to suppress inflammation is likely to be shadowed by the known adverse side effects. To circumvent this problem, our group is employing a nitrocellulose-based delivery approach for the local release of dexamethasone [92]. Local release of dexamethasone significantly reduced the reactivity of microglia and macrophages, as well as the expression of the inhibitory molecule CSPG 1 week postimplantation (Figure 6.6). Furthermore, such treatment greatly attenuated astroglial response and reduced neuronal loss in the vicinity of the electrode at 1 and 4 weeks postimplantation, implying that such an approach could potentially contribute to improving chronic neuronal recording.

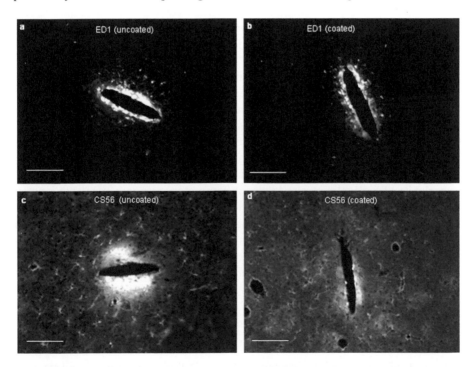

FIGURE 6.6 Representative images of horizontal brain sections inserted with control silicon electrode (a and c) and dexamethasone-coated electrode (b and d) 1 week postimplantation. ED1 stains for reactive microglia and macrophages (a and b) and CS56 stains chondroitin sulfate proteoglycans (c and d). Scale bar = 100 μm.

Unlike the synthetic glucocorticoid dexamethasone, α-MSH is endogenously expressed in cells of the immune system as well as in cells of the nervous and endocrine systems [93]. The neuropeptide has remarkable antiinflammatory properties, acting directly upon peripheral host cells to modulate the release of inflammatory substances and acting on receptors within the brain that drive descending antiinflammatory pathways [94]. Even though the strategy of delivering antiinflammatory agents can modulate the tissue response to electrodes, it presents the difficulty of sustaining the delivery over long implantation periods (months or even years). To complement this strategy, we proposed intrinsically modifying the electrode surface property by chemically immobilizing the α-MSH to the surface [95]. The peptide is grafted onto the surface using conventional silane chemistry. An *in vitro* cell study indicates that the immobilized peptide retained its biological property, making the electrode surface inherently antiinflammatory. In rats, the peptide-modified electrode reduced the microglial response at 1 and 4 weeks postimplantation, compared to the untreated control. As microglia are the frontline cells in direct contact with the electrode surface, their response is directly affected by the tethered α-MSH peptide. The astrocytic response was attenuated at the 4-week time point. It will be important to examine how long the tethered peptide will remain on the surface inside the tissue and whether it will be sufficient to modulate the gliosis response induced by "frustrated" phagocytosis as well as micromotion-induced mechanical strain. It is indisputable that to achieve maximum long-term improvement in tissue response, such an immobilization strategy should be applied in combination with other molecular interventions and serve as a foundation for a multilayered combinatorial treatment strategy.

Besides directly delivering antiinflammatory chemical drugs to influence cellular response, sustained release of antisense oligodeoxynucleotides (ODNs) against NF-κB mRNA would also be an effective strategy to diminish inflammation at the electrode implantation site. NF-κB antisense ODN is anticipated to selectively block proinflammatory gene expression, thus regulating cellular response at the mRNA level. Alternatively, strategies using antibodies against major initiators of gliosis, for example, TGF-β, could potentially mitigate the formation of glial scar. Given the complexity of molecules involved in the inflammatory response, many other potential molecular targets could be identified for therapeutic purposes. However, one has to be careful when employing such strategies, as increasing evidence has shown that many inflammatory molecular mediators in the CNS act like double-edged swords. Their effects do not occur in an all-or-none fashion; rather, they are concentration dependant. Therefore, it should be kept in mind that the goal is to contain any excessive activation rather than to remove all activity.

6.3.3   STRATEGY III: PROMOTING MIGRATION AND SURVIVAL OF NEURONS AND NEURITES

The electrode's main point of contact with the CNS is the neuron. Therefore, electrode performance is directly associated with the density of neurons and their proximity to the electrode sites. Theoretically, a recording electrode can detect action potentials from neurons within a radius of approximately 130 μm [96]. However, studies have suggested that the maximum recording distance is between 50 and 100 μm [97–100]. Accompanying brain tissue response to the implanted electrode is a reported neuronal cell loss. About a 40% decrease in neuronal cell bodies within a 100-μm radius of the electrode–tissue interface was observed at 2 and 4 weeks post-implantation [68]. The loss is mainly attributed to the production of neurotoxic factors such as proinflammatory cytokines and free oxygen radicals that are produced during persistent inflammation, rather than to direct mechanical trauma during electrode implantation [68]. This indicates that neuronal survival can be indirectly enhanced by attenuating the inflammatory response using the strategies suggested above.

A more active approach would be to directly promote neuronal survival and growth toward the electrode by releasing neurotrophic factors. A classical example is the study conducted by Kennedy [101] in which glass cone electrodes seeded with pieces of sciatic nerve recorded neural signals with signal to noise ratios that were 5 to 10 times better than those obtained with wire and silicon electrode arrays for over 12 months. It is likely that such improvement was related to the growth factors or chemoattractants released from the sciatic nerve. The neurotrophin family, a subgroup of the neurotrophic factors family, contains members such as NGF, brain-derived neurotrophic factor (BDNF), neurotrophin-3 (NT-3), and NT-4/5. These neurotrophins are widely expressed in nearly all neuronal populations in the CNS and peripheral nervous system (PNS) and are well known for their role in neuronal survival, process outgrowth, and regulation of synaptic plasticity [102]. Additionally, the chemotactic guidance of various migrating neurons is potentially mediated by the neurotrophins. In fact, delivery of growth factors has been proposed and is currently being studied as a potential therapeutic treatment for spinal cord injury [103]. A similar approach can be applied to electrodes implanted in the CNS: nourishing neurons with growth factors so that the number of recordable neurons will be increased. The challenges will be determining the parameters for delivery, specifically, the onset point, the dosage, the duration, and the delivery vehicle. Putatively, such strategic neurotrophin release could be an adjunctive component of therapeutics for achieving stable, long-term electrode performance.

Besides delivering exogenous growth factors, ideas of making the electrode surface more neural adhesive have been tested. *In vitro* studies have demonstrated that modifying the substrate surface with bioactive molecules such as laminin [104,105], the peptide fragment from laminin YIGSR [106], and cell adhesion molecule L-1 [107] significantly promoted neuron growth on the surface. However, the efficacy of these approaches *in vivo* is yet to be determined.

The other strategy to increase neuron density in the proximity of the electrode is to clear the path for neurites toward the interface. It is well known that the glial scar is highly inhibitory for neurite outgrowth, because of the presence of inhibitory components such as myelin-associated molecules (myeline-associated glycoprotein [MAG], Nogo, tenascin R) and CSPGs [87]. To encourage neurites to interact intimately with the electrode, this inhibitory environment has to be cleared. At least four possible approaches have been suggested [87], including (a) removing the cells that make inhibitory molecules, mostly astrocytes; (b) preventing the synthesis of inhibitory molecules, such as TGF-β; (c) blocking inhibitory molecules with antibodies that are specific to certain epitopes; and (d) degrading inhibitory molecules, such as using the bacterial enzyme chondroitinase ABC to digest away the glycosaminoglycan side chains of CSPG while leaving the core protein intact.

6.4 CONCLUSION

In this chapter, we have attempted to outline the differing molecular responses so that attempts to modulate tissue response to electrodes can be rationally pursued. Overall, the complex nature of CNS tissue response to chronic electrodes suggests that no single cellular or molecular targeting strategy will prove sufficient for achieving stable and long-term functional outcomes. Instead, a combinatorial strategy that is rationally designed and tailored, aided by a better understanding of underlying biological processes, will lead to cumulative improvements in interfacing technology that will one day be translated into real clinical applications.

ACKNOWLEDGMENTS

Funding support from National Institutes of Health, 1R01 DC06849 and 1R01 NS043486, is gratefully acknowledged (RVB). Special gratitude is expressed to Yinghui Zhong, George C. McConnell, and Thomas Schneider for their illuminating discussions and help in preparation of this manuscript.

REFERENCES

1. Spelman, F. A. The past, present, and future of cochlear prostheses. *IEEE Eng. Med. Biol. Mag.* 18, 27, 1999.
2. Lozano, A. M. et al. Deep brain stimulation for Parkinson's disease: Disrupting the disruption. *Lancet Neurol.* 1, 225, 2002.
3. Hochberg, L. R. et al. Neuronal ensemble control of prosthetic devices by a human with tetraplegia. *Nature* 442, 164, 2006.
4. Nicolelis, M. A. et al. Chronic, multisite, multielectrode recordings in macaque monkeys. *Proc. Natl. Acad. Sci. USA* 100, 11041, 2003.
5. Kennedy, P. R., Bakay, R. A., and Sharpe, S. M. The cone electrode: Ultrastructural studies following long-term recording in rat and monkey cortex. *Neurosci. Lett.* 142, 89, 1992.
6. Sachs, H. G. et al. Transscleral implantation and neurophysiological testing of subretinal polyimide film electrodes in the domestic pig in visual prosthesis development. *J. Neural. Eng.* 2, S57, 2005.
7. Drake, K. L. et al. Performance of planar multisite microprobes in recording extracellular single-unit intracortical activity. *IEEE Trans. Biomed. Eng.* 35, 719, 1998.
8. Campbell, P. K. et al. A silicon-based, three-dimensional neural interface: Manufacturing processes for an intracortical electrode array. *IEEE Trans. Biomed. Eng.* 38, 758, 1991.
9. Rousche, P. J. and Normann, R. A. Chronic recording capability of the Utah intracortical electrode array in cat sensory cortex. *J. Neurosci. Methods* 82, 1, 1998.
10. Polikov, V. S., Tresco, P. A., and Reichert, W. M. Response of brain tissue to chronically implanted neural electrodes. *J. Neurosci. Methods* 148, 1, 2005.

11. Wise, K. D. Silicon microsystems for neuroscience and neural prostheses. *IEEE Eng. Med. Biol. Mag.* 24, 22, 2005.
12. Rousche, P. J. et al. Flexible polyimide-based intracortical electrode arrays with bioactive capability. *IEEE Trans. Biomed. Eng.* 48, 361, 2001.
13. Szarowski, D. H. et al. Brain response to micro-machined silicon devices. *Brain Res.* 983, 23, 2003.
14. Cavaglia, M. et al. Regional variation in brain capillary density and vascular response to ischemia. *Brain Res.* 910, 81, 2001.
15. Bjornsson, C. S. et al. Effects of insertion conditions on tissue strain and vascular damage during neuroprosthetic device insertion. *J. Neural Eng.* 3, 196, 2006.
16. Klatzo, I. Pathophysiological aspects of brain edema. *Acta Neuropathol.* 72, 236, 1987.
17. Schilling, L. and Wahl, M. Mediators of cerebral edema. *Adv. Exp. Med. Biol.* 474, 123, 1999.
18. Nishino, A. et al. Thrombin may contribute to the pathophysiology of central nervous system injury. *J. Neurotrauma* 10, 167, 1993.
19. Pindon, A., Berry, M., and Hantai, D. Thrombomodulin as a new marker of lesion-induced astrogliosis: involvement of thrombin through the G-protein-coupled protease-activated receptor-1. *J. Neurosci.* 20, 2543, 2000.
20. Kubo, Y. et al. Thrombin inhibitor ameliorates secondary damage in rat brain injury: Suppression of inflammatory cells and vimentin-positive astrocytes. *J. Neurotrauma* 17, 163, 2000.
21. Xue, M. and Del Bigio, M. R. Acute tissue damage after injections of thrombin and plasmin into rat striatum. *Stroke* 32, 2164, 2001.
22. Xi, G., Reiser, G., and Keep, R. F. The role of thrombin and thrombin receptors in ischemic, hemorrhagic and traumatic brain injury: deleterious or protective? *J. Neurochemistry* 84, 3, 2003.
23. Ryu, J. et al. Thrombin induces NO release from cultured rat microglia via protein kinase C, mitogen-acivated protein kinase, and NF-kappa B. *J. Biol. Chem.* 275, 29955, 2000.
24. Möller, T., Hanisch, U.K., and Ransom, B. R. Thrombin-induced activation of cultured rodent microglia. *J. Neurochem.* 75, 1539, 2000.
25. Suo, Z. et al. Participation of protease-activated receptor-1 in thrombin-induced microglial activation. *J. Neurochem.* 80, 655, 2002.
26. Suo, Z. et al. Persistent protease-activated receptor 4 signaling mediates thrombin-induced microglial activation. *J. Biol. Chem.* 278, 31177, 2003.
27. Suo, Z., Citron, B. A., and Festoff, B. W. Thrombin: a potential proinflammatory mediator in neurotrauma and neurodegenerative disorders. *Curr. Drug Targets Inflamm. Allergy* 3, 103, 2004.
28. Nicole, O. et al. Activation of protease-activated receptor-1 triggers astrogliosis after brain injury. *J. Neurosci.* 25, 4319, 2005
29. Möller, T., Weinstein, J. R., and Hanisch, U. K. Activation of microglial cells by thrombin: past, present, and future. *Semin. Thromb. Hemost.* 32(Suppl. 1), 69, 2006.
30. Coughlin, S. R. Thrombin signaling and protease-activated receptors. *Nature* 407, 258, 2000.
31. Noorbakhsh, F. et al. Proteinase-activated receptors in the nervous system. *Nat. Rev. Neurosci.* 4, 981, 2003.
32. Beecher, K. L. et al. Thrombin receptor peptides induce shape change in neonatal murine astrocytes in culture. *J. Neurosci. Res.* 37, 108, 1994.

33. Ehrenreich, H. et al. Thrombin is a regulator of astrocyte endothelin-1. *Brain Res.* 600, 201, 1993.
34. Grabham, P. and Cunningham, D. D. Thrombin receptor activation stimulates astrocyte proliferation and reversal of stellation by distinct pathways: involvement of tyrosine phosphorylation. *J. Neurochem.* 64, 583, 1995.
35. Wang, H. et al. Thrombin (PAR-1)-induced proliferation in astrocytes via MAPK involves multiple signaling pathways. *Am. J. Physiol. Cell Physiol.* 283, C1351, 2002.
36. Wang, H., Ubl, J. J., and Reiser, G. The four subtypes of protease-activated receptors, co-expressed in rat astrocytes, evoke different physiological signaling. *Glia* 37, 53, 2002.
37. Xue, M. and Del Bigio, M. R. Acute tissue damage after injections of thrombin and plasmin into rat striatum. *Stroke* 32, 2164, 2001.
38. Fitch, M. T. and Silver, J. Activated macrophages and the blood-brain barrier: inflammation after CNS injury leads to increases in putative inhibitory molecules. *Exp. Neurol.* 148, 587, 1997.
39. McConnell, G. C. and Bellamkonda, R. V. Extraction force and cortical tissue reaction of silicon microelectrode arrays implanted in the rat brain. Unpublished manuscript.
40. Krueger, S. et al. Three morphologically distinct types of interface develop between adult host and fetal brain transplants: Implications for scar formation in the adult central nervous system. *J. Comp. Neurol.* 249, 103, 1986.
41. Nathaniel, E. J. H. and Nathaniel, D. R. The reactive astrocyte. In Federoff, S., ed., *Advances in Cellular Neurobiology.* Orlando, FL: Academic Press, 1981. pp. 249–301.
42. Chen, Y. and Swanson, R. A. Astrocytes and brain injury. *J. Cereb. Blood Flow Metab.* 23, 137, 2003.
43. Wang, C. X. and Shuaib, A. Involvement of inflammatory cytokines in central nervous system injury. *Prog. Neurobiol.* 67, 161, 2002.
44. Aloisi, F. Immune function of microglia. *Glia* 36, 165, 2001.
45. Ghirnikar, R. S., Lee, Y. L., and Eng, L. F. Inflammation in traumatic brain injury: Role of cytokines and chemokines. *Neurochem. Res.* 23, 329, 1998.
46. Streit, W. J. et al. Cytokine mRNA profiles in contused spinal cord and axotomized facial nucleus suggest a beneficial role for inflammation and gliosis. *Exp. Neurol.* 152, 74, 1998.
47. Louis, J. C. et al. CNTF protection of oligodendrocytes against natural and tumor necrosis factor-induced death. *Science* 259, 689, 1993.
48. Downen, M. et al. Neuronal death in cytokine-activated primary human brain cell culture: Role of tumor necrosis factor-alpha. *Glia* 28, 114, 1999.
49. Scherbel, U., Raghupathi, R., and Nakamura, M. Differential acute and chronic responses of tumor necrosis factor-deficient mice to experimental brain injury chemical messengers. *Proc. Natl. Acad. Sci. USA* 96, 8721, 1999.
50. Baird, A. and Gage, F. H. Cytokine regulation of nerve growth factor-mediated cholinergic neurotrophic activity synthesized by astrocytes and fibroblast. *J. Neurochem.* 59, 919, 1992.
51. Vitkovic, L., Bockaert, J., and Jacque, C. "Inflammatory" cytokines: Neuromodulators in normal brain? *J. Neurochem.* 74, 457, 2000.
52. Herx, L. M., Rivest, S., and Yong, V. W. Central nervous system-initiated inflammation and neurotrophin in trauma: IL-1 beta is required for the production of ciliary neurotrophis factor. *J. Immunol.* 165, 2232, 2000.

53. Herx, L. M. and Yong, V. W. Interleukin-1 beta is required for the early evolution of reactive astrogliosis following CNS lesion. *J. Neuropathol. Exp. Neurol.* 60, 961, 2001.
54. Allan, S. M., Tyrrell, P. J., and Rothwell, N. J. Interleukin-1 and neuronal injury. *Nat. Rev. Immunol.* 5, 629, 2005.
55. Basu, A., Krady, J. K., and Levison, S. W. Interleukin-1: A master regulator of neuro-inflammation. *J. Neurosci. Res.* 78, 151, 2004.
56. John, G. R., Lee, S. C., and Brosnan, C. F. Cytokines: Powerful regulation of glial cell activation. *Neuroscientist* 9, 10, 2003.
57. John, G. R. et al. IL-1-regulated responses in astrocytes: Relevance to injury and recorvery. *Glia* 49, 161, 2005.
58. Shrikant, P. and Benveniste, E. N. The central nervous system as an immunocompetent organ: role of glia cells in antigen presentation. *J. Immunol.* 157, 1819, 1996.
59. Selmaj, K. W. et al. Proliferation of astrocytes *in vitro* in response to cytokines. A primary role for tumor necrosis factor. *J. Immunol.* 144, 129, 1990.
60. Unsicker, K. and Strelau, J. Functions of transforming growth factor-beta isoforms in the nervous system. Cues based on localization and experimental *in vitro* and *in vivo* evidence. *Eur. J. Biochem.* 26, 6972, 2000.
61. Benveniste, E. N., Nguyen, V. T., and O'Keefe, G. M. Immunological aspects of microglia: relevance to Alzheimer's disease. *Neurochem. Int.* 39, 381, 2001.
62. Logan, A. et al. Effects of transforming growth factor beta 1 on scar production in the injured central nervous system of the rat. *Eur. J. Neurosci.* 6, 355, 1994.
63. Yong, V. W. et al. Matrix metalloproteinases and diseases of the CNS. *Trends Neurosci.* 21, 75, 1998.
64. Wang, X. et al. Mechanical injury in rat cortical cultures activates MAPK signaling pathways and induces secretion of matrix metalloproteinase-2 and –9. *J. Cereb. Blood Flow Metab.* 21, s264, 2001.
65. Morita-Fujimura, Y. et al. Overexpression of copper and zinc superoxide dismutase in transgenic mice prevents the induction and activation of matrix metalloproteinases after cold injury induced brain trauma. *J. Cereb. Blood Flow Metab.* 20, 130, 1999.
66. Denninger, J. W. and Marletta, M. A. Guanylate cyclase and the NO/cGMP signaling pathway. *Biochem. Biophys. Acta* 1411, 334, 1999.
67. Hewett, S. J., Csernansky, C. A., and Choi, D. W. Selective potentiation of NMDA-induced neuronal injury following induction of astrocytic iNOS. *Neuron* 13, 487, 1994.
68. Biran, R., Martin, D. C., and Tresco, P. A. Neuronal cell loss accompanies the brain tissue response to chronically implanted silicon microelectrode arrays. *Exp. Neurol.* 195, 115, 2005.
69. Pearson, G. L., Read, W. T., and Feldman, W. L. Deformation and fracture of small silicon crystals. *Acta. Metall.* 5, 181, 1957.
70. Ommaya, A. K. Mechanical properties of tissues of the nervous system. *J. Biomech.* 1, 127, 1967.
71. Britt, R. H. and Rossi, G. T. Quantitative analysis of methods for reducing physiological brain pulsations. *J. Neurosci. Methods* 6, 219, 1982.
72. Fee, M. S. Active stabilization of electrodes for intracellular recording in awake behaving animals. *Neuron* 27, 461, 2000.
73. Goldstein, S. R. and Salcman, M. Mechanical factors in the design of chronic recording intracortical microelectrodes. *IEEE Trans. Biomed. Eng.* 20, 260, 1973.
74. Gilletti, A. and Muthuswamy, J. Brain micromotion around implants in the rodent somatosensory cortex. *J. Neural Eng.* 3, 189, 2006.

75. Lee, H. et al. Biomechanical analysis of silicon microelectrode-induced strain in the brain. *J. Neural Eng.* 2, 81, 2005.
76. Subbaroyan, J., Martin, D. C., and Kipke, D. R. A finite-element model of the mechanical effects of implantable microelectrodes in the cerebral cortex. *J. Neural Eng.*, 2, 103, 2005.
77. Ostrow, L. W. and Sachs, F. Mechanosensation and endothelin in astrocytes: Hypothetical roles in CNS pathophysiology. *Brain Res. Rev.* 48, 488, 2005.
78. Floyd, C. L., Gorin, F. A., and Lyeth, B. G. Mechanical strain injury increases intracellular sodium and reverses Na^+/Ca^{2+} exchange in cortical astrocytes. *Glia* 51, 35, 2005.
79. Floyd, C. L. et al. Traumatic injury of cultured astrocytes alters inositol (1,4,5)-trisphophate-medicated signaling. *Glia* 33, 12, 2001.
80. Neary, J. T. et al. Activation of extracellular signal-regulated kinase by stretch-induced injury in astrocytes involves extracellular ATP and P2 purinergic receptors. *J. Neurosci.* 23, 2348, 2003.
81. Rzigalinski, B. A. et al. Effect of Ca^{2+} on *in vitro* astrocyte injury. *J. Neurochem.* 68, 289, 1997.
82. Lamb, R. G. et al. Alterations in phosphatidylcholine metabolism of stretch-injured cultured rat astrocytes. *J. Neurochem.* 68, 1904, 1997.
83. Willoughby, K. A. et al. S100B protein is released by *in vitro* trauma and reduces delayed neuronal injury. *J. Neurochem.* 91, 1284, 2004.
84. Ahmed, S. M. et al. Stretch-induced injury alters mitochondrial membrane potential and cellular ATP in cultured astrocytes and neurons. *J. Neurochem.* 74, 1951, 2000.
85. Tavalin S. J., Ellis E. F., and Satin L. S. Mechanical perturbation of cultured cortical neurons reveals a stretch-induced delayed depolarization. *J. Neurophysiol.* 74, 2767, 1995.
86. Zhong, Y. and Bellamkonda, R. V. Response of glial cells to cyclic strain. In preparation, 2007.
87. Fawcett, J. W. and Asher, R. A. The glial scar and central nervous system repair. *Brain Res. Bull.* 49, 377, 1999.
88. Wu, C. C. et al. Selective inhibition of protease-activated receptor 4-dependent platelet activation by YD-3. *Thromb. Haemost.* 87, 1026, 2002.
89. Covic, L. et al. Pepducin-based intervention of thrombin-receptor signaling and systemic platelet activation. *Nat. Med.* 8, 1161, 2002.
90. Ichiyama, T. et al. Systemically administered α-melanocyte-stimulating peptide inhibit NF-κB activation in experimental brain inflammation. *Brain Res.* 836, 31, 1991.
91. Spataro, L. et al. Dexamethasone treatment reduces astroglia responses to inserted neuroprosthetic devices in rat neocortex. *Exp. Neurol.* 194, 289, 2005.
92. Zhong, Y. and Bellamkonda, R. V. Dexamethasone coated neural probes elicit attenuated inflammatory response and neuronal loss compared to uncoated neural probes. *Brain Res.* 1148, 15, 2007.
93. Rajora, N. et al. α-MSH production, receptors, and influence on neopterin in a human monocyte/macrophage cell line. *J. Leukoc. Biol.* 59, 248, 1996.
94. Lipton, J. M. et al. Mechanisms of anti-inflammatory action of α-MSH peptides. *Ann. NY Acad. Sci.* 885, 173, 1999.
95. He, W. et al. A novel anti-inflammatory surface for neural electrodes. *Adv. Nat.* (Accepted, 2007.)
96. Eaton, K. P. and Henriquez, C. S. Confounded spikes generated by synchrony within neural tissue models. *Neurocomputing* 65, 851, 2005.
97. Mountcastle, V. B. Modality and topographic properties of single neurons of cat's somatic sensory cortex. *J. Neurophysiol.* 20, 408, 1957.

98. Rall, W. Electrophysiology of a dendritic neuron model. *Biophy. J.* 2(2 Pt 2), 145, 1962.
99. Rosenthal, F. Extracellular potential fields of single PT-neurons. *Brain Res.* 36, 251, 1972.
100. Henze, D. A. et al. Intracellular features predicted by extracellular recordings in the hippocampus *in vivo*. *J. Neurophysiol.* 84, 390, 2000.
101. Kennedy, P. R. The cone electrode—A long-term electrode that records from neurites grown onto its recording surface. *J. Neurosci. Methods* 29, 181, 1989.
102. Lessmann, V., Gottmann, K., and Malcangio, M. Neurotrophin secretion: Current facts and future prospects. *Prog. Neurobiol.* 69, 341, 2003.
103. Thuret, S., Moon, L. D. F., and Gage, F. H. Therapeutic interventions after spinal cord injury. *Nat. Rev. Neurosci.* 7, 628, 2006.
104. He, W. and Bellamkonda, R. V. Nanoscale neuro-integrative coatings for neural implants. *Biomaterials* 26, 2983, 2005.
105. Ignatius, M. J. et al. Bioactive surface coatings for nanoscale instruments: effects on CNS neurons. *J. Biomed. Mater. Res.* 40, 264, 1998.
106. Cui, X. Y. et al. Surface modification of neural recording electrodes with conducting polymer/biomolecule blends. *J. Biomed. Mater. Res.* 56, 261, 2001.
107. Azemi, E. et al. Improving the biocompatibility of neural probes by surface immobilization of L1. Poster presented at Neural Interfaces Workshop, Bethesda, MD, August 21–23, 2006.

7 Soft, Fuzzy, and Bioactive Conducting Polymers for Improving the Chronic Performance of Neural Prosthetic Devices

Dong-Hwan Kim, Sarah Richardson-Burns,
Laura Povlich, Mohammad Reza Abidian,
Sarah Spanninga, Jeffrey L. Hendricks,
and David C. Martin

CONTENTS

7.1 INTRODUCTION

Microfabricated electrodes for stimulating and recording signals from individual neurons have facilitated direct electrical connections with living tissue. While these devices have worked reasonably well in acute applications, chronically implanted electrodes have had more limited success [1,2]. To improve the long-term integration of these devices, coatings have been developed to accommodate the differences in mechanical properties, bioactivity, and mechanisms of charge transport between the engineered electronic device and living cells [3–10]. Conducting polymers can be directly deposited onto electrode surfaces with precisely controlled morphologies. The coatings lower the impedance of the electrodes and provide a mechanical buffer between the hard device and the soft tissue. These coatings can be tailored to incorporate and deliver pharmacological agents such as antiinflammatory drugs and neurotrophic factors. *In vivo* studies to date have shown that these coatings improve the long-term recording performance of cortical electrodes [11].

In this review we first discuss the development of neural prosthetic devices, including the history of their development, issues associated with the electrode–tissue interface, inflammation and neural loss in the tissue near the electrode surface, the mechanical property differences between the probe and the tissue, the geometry of the probe, and materials used to modify the electrode surface. We then discuss the design of materials for the electrode–tissue interface to help these probes function more effectively over the long term. These materials are intended to improve device performance by creating a mechanically compliant (soft), high-surface-area (fuzzy), low-impedance electrode–tissue interface that can have controlled biological functionality. We conclude by describing the results of work to date that have focused on the design, synthesis, and characterization of electrode interface materials, with particular attention to the use of conducting polymers that have been shown to significantly improve the electrical properties at these interfaces.

7.2   OVERVIEW OF NEURAL PROSTHETIC DEVICES

Over the past 50 years, many different types of neural prosthetic devices have been used to record and stimulate neural signals in the central nervous system (CNS) and peripheral nervous system (PNS) [12,13]. Implanted microelectrodes of various designs, including microwires [14,15] and micromachined electrodes (Utah electrodes and Michigan electrodes) are currently being used for the treatment of deafness (cochlear implants) [16] and Parkinson's disease (deep brain stimulation) [17,18].

The elastic modulus of silicon is near 100 GPa, whereas that of brain is on the order of 100 kPa, akin to Jell-O [19,20]. This corresponds to a stiffness variation of approximately six orders of magnitude. The mismatch of stiffness at the tissue–device interface has the potential to create large interfacial strains during the lifetime of chronic implants. To avoid sharp interfaces between materials, it is necessary to provide a gradient of mechanical properties (Figure 7.1a). This interfacial mechanical mismatch has been addressed by applying deposition of electrospun nanofibers of polymer on the microelectrode arrays [3], and genetically engineered protein polymer materials have been applied to the devices to mediate the mechanical differences between the device and brain tissue with structural stability of natural silk with responsive properties (Figure 7.1b) [21]. Flexible polyimide-based microelectrode arrays are being fabricated that minimize the stress resulting from

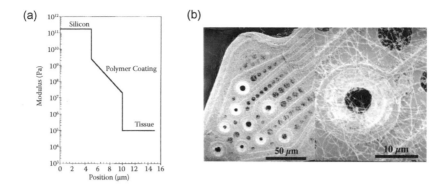

FIGURE 7.1   (a) Schematic of a modulus gradient in a polymer film that could help mediate the large differences in mechanical properties between a stiff prosthetic device and soft tissue [3]. (b) Scanning electron microscope images of nanofibrous coatings of an SLPF containing NGF deposited onto the surface of a sieve probe. The polymer-coated devices showed significantly improved neuronal ingrowth and long-term performance [21]. (Used with permission).

micromotion of stiff probes [22]. Recently, the Kipke group developed a cortical probe design that has the electrodes held off to the side of the main probe body [23] to float the electrode on a thin supporting member. The thicker, main part of the probe would still be stiff enough to facilitate insertion into tissue. Histological studies have shown a significant decrease in the amount of inflammation seen around the smaller lateral shank [23]. Although this design has not yet been implemented on functional probes, these results support the idea that electrodes with reduced stiffness increase the integration of the electrode into the tissue.

To understand the effects of the device design and the insertion method on cellular sheath development, a study of chronically implanted, single-shanked, chisel-tipped microelectrodes with trapezoid and square cross sections as well as blade-type, single-shanked microelectrodes from the Center for Neural Communication Technology (CNCT) (University of Michigan) was performed. These studies demonstrated that long-term reactivity is essentially independent of electrode size, shape, surface texture, and insertion method [24,25]. Clinical applications of neuroprosthetic devices could, however, benefit from miniaturization of these devices since the volume of surrounding tissue associated with early reactive response is associated with the amount of tissue damage during insertion.

During the insertion of electrodes into the brain, rupturing of blood vessels and damage to surrounding tissue is inevitable. For example, a typical Michigan probe measuring 15 μm × 60 μm × 2 mm will displace 0.0018 mm^3 of cortical tissue, resulting in displacement of about 50 neurons and 400,000 synapses, assuming that the human cortex contains approximately 30,000 neurons and 0.24 billion synapses in a cubic millimeter [26]. The mechanical trauma of electrode insertion initiates the wound healing process and subsequent cellular encapsulation around the implant, leading to decreased signals and eventual failure of communication between neurons and electrodes. It is hypothesized that the primary role of the glial scar is to separate the damaged tissue from the rest of the body to maintain the blood–brain barrier and prevent lymphocyte infiltration [27]. Szarowski et al. showed that this response includes an early reactive component observed immediately upon device insertion and a sustained reactive component that develops with time and is maintained as long as the implant is present [24].

The sustained reactive response, a major source of failure in chronically implanted electrodes, is associated with the characteristics of the surface materials of the implants. Sapphire [28] and alumina (Al_2O_3) [29] for the substrate of micromachined electrodes and polyesterimide-coated gold wire [30] have been used to improve performance of the electrodes. Electrostatic layer-by-layer (LbL) self-assembly techniques have been used for the surface modification of neural electrodes [31]. They used alternating polyelectrolytes, either polyethyleneimine or chitosan, and proteins, either laminin or gelatin, to fabricate multilayer films on silicon wafers. Huber et al. suggested other techniques such as adsorption, covalent coupling, and electrochemical polymerization that could be used to coat laminin-derived peptides on glassy carbon surfaces [32]. Silk-like polymers containing fibronectin fragments (SLPF) and nonapeptide (CDPGYIGSR) were immobilized into polypyrrole (PPy)-conducting polymers to enhance electric transportation and adhesion of neurons [4]. Microcontact printing, a new technique of chemically and molecularly patterning surfaces on a submicrometer scale, was also used to modify silicon substrates with polylysine [33]. Other studies have been performed with coatings containing living cells designed to better integrate the devices with living tissue. For example, Martin and Tresco implanted electrodes coated with olfactory ensheathing cells derived from the olfactory bulbs of adult Fischer 344 rats into the cerebral cortex for a period of 2 weeks. It was found that the reactivity surrounding cellular-coated electrodes was significantly reduced compared to the controls [34].

7.3 RECORDING SITE–NEURON INTERFACE

Signal transmission along neurons is the result of ionic currents generated by movement of ions via specific ion channels in the cellular membrane. When a neuron generates an action potential, the current flows within the cell and leaks though regions of membrane. The flow of current across the membrane generates a complex potential field around the neuron. The magnitude and orientation of this field depend on the size, geometry, and location of the cell and the time course of depolarization [35]. The time course of this potential difference is what gives rise to the traditional extracellular action potential. The action potentials are transient waveforms with a typical duration of 1 msec.

Extracellular recording methods have been used to obtain data about the properties of CNS structures, including the mapping of field potentials from a single neuron to the excitability of CNS dendrites and studying behaviorally related discharge patterns of CNS neurons [36]. In addition, extracellular recording makes it possible to monitor the activity of multiple neurons in the vicinity of the electrode. Although conduction at the electrode–electrolyte interface is not yet fully understood, it involves a capacitive mechanism (charging and discharging of the electrode double layer) and a Faradaic mechanism (chemical oxidation and reduction). In general, the diffusion and recombination of ionic species from the liquid is a nonlinear

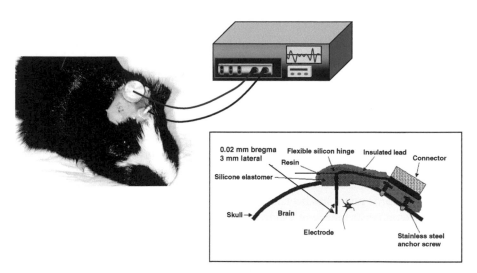

FIGURE 7.2 Michigan electrodes chronically implanted into auditory cortex of guinea pig and extracellular recording of neural signals.

charge-transfer process [37,38]. Figure 7.2 shows Michigan electrodes chronically implanted into the auditory cortex of a guinea pig. Five to ten seconds of continuous neural recordings are collected using a National Instruments data acquisition card coupled to a Plexon multichannel acquisition processor running RASPUTIN. The auditory cortex recordings are driven by 200-ms noise bursts presented 2 per second using a digital to analog converter (Tucker-Davis Technologies). The round cap on the guinea pig's head covers and protects the chronic electrodes. The electrode connector is placed under the cap. The rod-shaped screw next to the cap of the electrodes functions as a ground when the neural spikes are recorded (Figure 7.2).

It is known that the amplitude of the extracellular spike is much smaller than the corresponding intracellular spike, because the magnitude of the extracellular current is smaller to begin with and diminishes rapidly as a function of distance from the cell [39]. To record extracellular potentials with the greatest signal and spatial selectivity, electrodes are required to have recording sites with dimensions as small as 20 μm, and it is highly recommended that they be placed as close as possible to the firing neuron. Henze et al. claims that at distances >50 μm from the soma, individual spikes could be recognized with extracellular recordings (the reliability of unit separation for these signals is decreased significantly). At distances >140 μm from the soma, extracellular spikes could not be distinguished from the background noise level [40].

7.4 BIOMATERIAL COATINGS ON IMPLANTED ELECTRODES

In considering the design of materials for the interface between living tissue and an engineered electronic device, it is necessary to consider the dramatic differences in structure and properties between these two systems. Living tissue is wet, whereas biomedical devices are usually solids. Tissue is usually quite soft, whereas devices are often hard. Tissue conducts electron charge by ionic transport, whereas devices conduct charge with electrons and holes. Any material that functions at the interface between the biotic tissue and abiotic device will therefore need to take these large variations of behavior into account. Conducting polymers are particularly attractive for this purpose because they have mechanical properties between those of the metal electrode and the tissue and are able to facilitate charge transport with various cationic and anionic species. They can also be complexed with a variety of biologically active counter-ions. The ultimate limits of conducting polymers for use in biosensors were discussed by Goepel, who described the concept of a "molecular wire" that could facilitate signal transduction from biological tissue to an electrically conducting metal substrate [41].

Our research group has been developing materials and processes for improving the long-term performance of electronic biomedical devices, with particular interest in the microfabricated, implantable cortical probes that have been developed at the University of Michigan. We began by focusing on genetically engineered protein polymers that combine amino acid sequences from structural proteins such as silk and elastin with binding sequences from matrix proteins such as fibronectin and laminin. It was found that continuous films of these polymers were not as useful as filamentous structures that could be created by a novel process called electrospinning. Electrospinning creates high-surface-area structures that promote cellular ingrowth (Figure 7.3). This open microstructure also made it possible to maintain carrier transport to the device [42] and provided a mechanically compliant

(a)   (b)

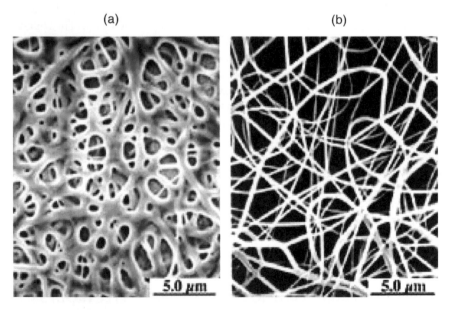

FIGURE 7.3   SEM images of electrospun filaments of protein polymers with tailored morphologies. The diameters of the filaments created by electrospinning can be controlled by changing the solution concentration, electric field, and distance from the substrate. The filaments deposited from closer to the probe are flatter and form a dense mat (a), while the filaments deposited from farther away are rounded and open (b) [43]. (Used with permission.)

surface for interfacing with the soft tissue of the brain [3]. Electrospun filaments can also be loaded with biologically active molecules for controlled release into tissue [10,43–45].

However, one important limitation of the electrospinning process is that the polymer coats the entire probe with fibers. Indeed, any type of nonconducting polymer coating will not be useful for these microfabricated electronic devices, since the electrodes need to remain extremely sensitive for single-cell recording. For improving the electrical transport properties of the probe, and for controlling the composition of the material in the vicinity of the recording site, we have since developed schemes that have allowed them to exert more direct control on the environment closer to the active sites on the probe. Specifically, we have investigated conducting polymers that can be electrochemically deposited directly on the electrodes [4].

A key requirement for a polymer to become intrinsically conductive is the molecular orbital overlap to allow formation of a delocalized molecular wave function. Conducting polymers have a delocalized, extended π-bonded system of electrons resulting from the conjugation of alternating single and double bonds along the molecular backbone. Applications of conducting polymers include analytical devices and biosensors [46–51]. They have afforded new surface modification strategies for functionalization of conventional metal electrodes for creating pharmacological and toxicological biochemical sensors. Conducting polymers have also been utilized for light-emitting diodes [52], photovoltaic solar cells [53], lightweight batteries [54], antistatic coatings, and electrochromic devices [55].

Among the currently available inherently conducting polymers (Figure 7.4a), PPy (Figure 7.4b) has been extensively investigated because it is highly conductive, easily oxidized, and electropolymerizable from water. The deposited PPy film is reasonably stable and adherent to the electrode [56,57]. The biocompatibility of PPy has been investigated *in vitro* and *in vivo* [5,58,59]. PPy has been used to modify bioelectronic devices [4,60–62]P and to create biosensors by immobilization of biological elements [63] including enzymes [64,65], antibodies [66,67], and DNA [67].

Electrochemical polymerization is the preferred technique for the synthesis of conducting polymers for coatings on biomedical devices, since it is reproducibly controlled, provides the highest conductivities, and deposits the polymer only on specified areas. As shown in the reaction scheme (Figure 7.4b,c), the pyrrole and ethylenedioxythiophene (EDOT) monomer are electrochemically polymerized into the polymer PPy and poly(3,4-ethylenedioxythiophene) (PEDOT). The oxidation of monomers by the application of current to the monomer solution forms cationic radicals that eventually lead to polymerization. Electrochemical polymerization of conducting polymers is performed by application of constant current (galvanostatic), constant potential (potentiostatic), or potential scanning or sweeping methods. The total electrical charge passing through the electrode drives this stoichiometric reaction, leading to coatings of precisely defined composition, thickness, and microstructure. The polymerization reaction is typically performed in a three-electrode cell including a counter electrode (anode), a reference electrode, and a working electrode (cathode). Metals such as gold, titanium, platinum, and chromium are often used for the counter electrode [68].

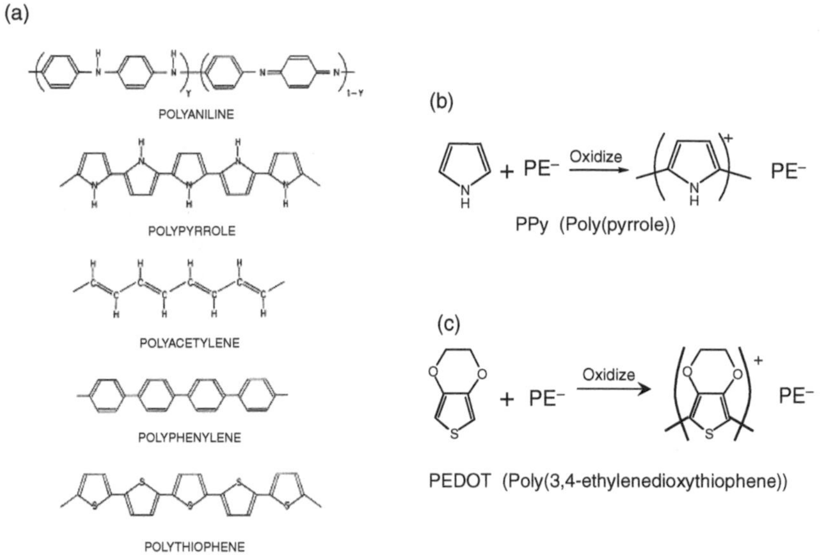

(a)

POLYANILINE

(b)

POLYPYRROLE

PPy (Poly(pyrrole))

POLYACETYLENE

(c)

POLYPHENYLENE

PEDOT (Poly(3,4-ethylenedioxythiophene))

POLYTHIOPHENE

FIGURE 7.4 Structures of conducting polymers (a) and schematics of the oxidative elec-trochemical polymerization of polypyrrole (b) and poly(3,4-ethylenedioxythiophene) (c). The polymers have a net positive charge in their electrically active state and are typically complexed with a polyelectrolyte that can be composed of various ionically charged species or biological molecules such as proteins or peptides.

Cui et al. established that electrochemical polymerization can be used to deposit coatings of electrically conducting polymers such as PPy directly onto metal neural electrode sites [4] (Figure 7.5). Cui et al. also found that the impedance magnitude can be tailored by changing the applied current and that the lowest impedance could be obtained with a PPy thickness of 10 to 12 µm, depending on the dopant. This optimum thickness is apparently related to the fact that the impedance is profoundly influenced by charge carrier transport across the electrode–tissue interface. Thinner films are flat, and the effective surface area at the interface between the electrode sites and neural tissue is relatively small. During growth, the surface roughens, increasing the effective surface area of the interface [4] (Figure 7.6).

Neural electrodes are often used for chronic applications that demand the chemical and electrical stability of the implanted materials. Yamato et al. found that PPy has limited electrical response under cyclic voltammetry because of its poorly defined chemical structure [69]. Heywang and Jonas reported the synthesis of a variety of poly(alkylenedioxythiophenes), including PEDOT [70]. PEDOT exhibits improved conductivity and thermal stability because the dioxylethylene bridging groups across the 3 and 4 positions of the EDOT monomer block the possibility of α-β' coupling. Yamato et al. reported that PEDOT–polystyrenesulfonate (PSS) was more chemically stable than PPy–PSS [69]. After polarization for 16 h at 400 mV and pH 7.5, only 5% of the original electrochemical activity of PPy–PSS remained, whereas PEDOT–PSS retained 89% of its original activity [69]. Along with PPy,

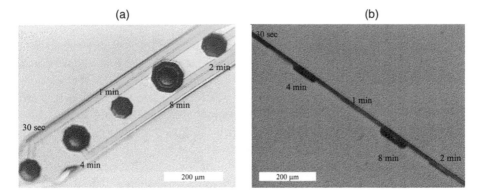

FIGURE 7.5 Optical micrographs of electropolymerization of PPy on a neural prosthetic device seen from the above (a) and from the side (b). The thickness of the conducting polymer film can be precisely tailored by controlling the solution concentration, current density, and time of reaction.

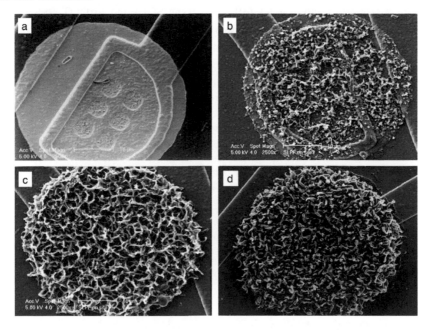

FIGURE 7.6 SEM images of PPy–SLPF-coated neural electrodes. From (a to d), the deposition time was increased at constant current, so the total amount of charge that passed through during the deposition increased. (a) Bare gold, (b) 1 mC, (c) 4 mC, (d) 10 mC. Scale bar = 10 μm. (Adapted from Cui, X. Y. et al., *J. Biomed. Mater. Res.* 56(2), 261, 2001. With permission.)

the PEDOT family of polymers has also been intensively investigated for neural prosthetic applications [71–74]. Xiao et al. also investigated PEDOT derivatives such as poly(hydroxymethylated-3,4-ethylenedioxylthiophene) (PEDOT-MeOH) and PEDOT doped with its derivative sulfonatoalkoxyethlyenedioxythiophene (S-

EDOT) to overcome the low water solubility of EDOT that results in limited immobilization of biologic agents into PEDOT [72–74].

It has been reported that the effective surface area of the neural electrode site is crucial in determining its electrical properties and in providing a gradient of mechanical properties that promotes better integration with cells [5,71]. Yang et al. explored a number of methods to create features of well-defined size and high surface area of conducting polymers using templating techniques. A variety of nanostructured conducting polymer morphologies have been fabricated as coatings including the nanomushroom, nanohair, and nanopore structures [6,7,75,76] (Figure 7.7). Highly ordered microporous PEDOT-conducting polymer coatings fabricated using polystyrene latex sphere templates had the lowest impedance value (~10 kΩ), with 300-nm scale-polystyrene spheres [6]. These spheres had the smallest diameters that still retained relatively high order in the film, creating an open-cell foam that facilitates charge transport. The oxidation potential for microporous PEDOT–LiClO$_{44B}$ was -0.7 V versus a saturated calomel electrode (SCE), which is significantly lower than that of microporous PPy–LiClO$_4$ (−0.4 V versus SCE) and PEDOT–PSS (-0.4 V versus SCE) and PPy–PSS (0.1 V versus SCE) [71]. These lower oxidation potentials are critical in preventing degradation from biological reducing agents in living tissue [76]. Yang et al. also developed PEDOT films that can be electrochemically deposited through self-assembled nonionic surfactant films on surfaces of neural electrodes [77], following methods reported by Hulvat and Stupp [78].

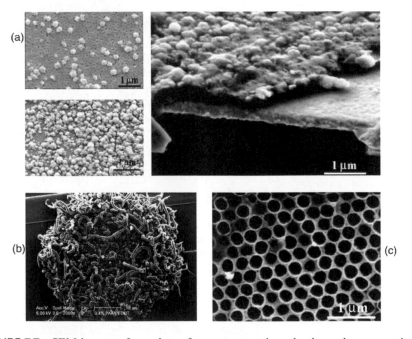

FIGURE 7.7 SEM images of a variety of nanostructured conducting polymer morphologies fabricated using templating techniques including nanomushroom (a); nanohair (b); and nanopore (c) structures.

The performance of polymer-coated electronic devices involves the transfer of charge through a variety of interfaces and through materials with different species that are the mobile charge carriers. In metal electrodes charge transport involves the motion of electrons, whereas in conducting polymers the charge is predominantly transported by local positive charges that move along the molecular backbone (holes or polarons) (Figure 7.8). In water-laden tissue, charge transport involves the motion of positively charged cations or negatively charged anions (Figure 7.8). Transfer reactions involving exchange of charge between these various carriers occur at the metal–polymer and polymer–tissue interfaces [38]. It is reasonable to expect that the detailed composition and microstructure of these various interfaces will be critical for the performance of electronic biomedical devices implanted in living, ionically conducting tissue. In particular, it is anticipated that increasing the effective

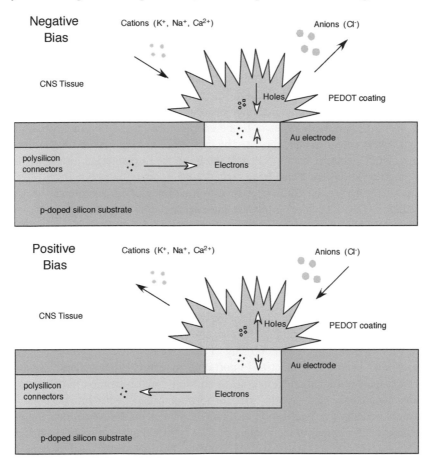

FIGURE 7.8 Schematic diagram showing the mechanisms of charge transport near the electrode–tissue interface under different potential bias. With negative bias, there is electron transport through the polysilicon connectors and metal electrode. There is a reverse flow of holes through the conducting PEDOT coating. Cations such as potassium, sodium, and calcium ions are driven toward the polymer, and anions such as chlorine are driven out into the solution. In positive bias, the carriers are the same, but the direction of travel is reversed.

surface area will allow for many more opportunities for such charge transfer to occur, significantly lowering the overall impedance of the device.

Electrochemical impedance spectroscopy measurements have shown that the charge transport characteristics of the polymer coatings are extremely sensitive to the surface morphology. The rough, fuzzy structure of the electrochemically deposited conducting polymer leads to a significant (~2 orders of magnitude) decrease in the sample impedance. Quantitative comparisons of the impedance response as a function of temporal frequency have shown correlations with the surface topology of the coatings as measured by atomic force microscopy (AFM) [79]. It has also been shown that the microstructure and electrical properties of the coatings correlate with their mechanical behavior, as measured by nanoindentation [80]. The fuzzy, low-impedance coatings also have the lowest effective modulus, consistent with a more open physical architecture.

PEDOT-coated microelectrodes have recently been studied *in vivo*. The results showed high-quality spike recordings at 6 weeks postimplant from PEDOT-coated electrodes. The signal-to-noise ratios for the PEDOT-coated sites were higher than the uncoated iridium electrode sites, and there was a 20% increase in the average number of units recorded per site [11].

Polymer processing methods have proven to be relatively easy to scale up and have been used on many types of probe geometries. In addition to chronic probes being actively designed and fabricated by the CNCT at Michigan, multishank neural probes (Utah electrode arrays [UEAs]) intended for implantation into feline dorsal root ganglia (DRG) have also been coated with conducting polymers [81,82]. The conducting polymers have also been investigated for applications such as deep brain stimulation (DBS), cochlear implants, and pacemakers. In the future it may also be possible to adapt similar approaches to other devices for which it is important to maintain electrical contact with living tissue such as electroencephalogram skin sensors or glucose sensors.

7.4.1   Incorporation of Biologic Species into Conducting Polymers

A biosensor is a device that has a biological sensing element capable of producing a signal that is a function of the concentration of a specific chemical or set of chemicals. Specifically, the monitoring of metabolites such as glucose, urea, cholesterol, and lactate is of central importance in clinical diagnostics. Conducting polymers have attracted much interest as suitable immobilization matrices for biosensors because the electrochemical incorporation of biospecies into the conducting polymers permits the localization of these molecules on the electrodes of any size or geometry. A number of techniques such as physical adsorption [83], entrapment [84], cross-linking [85], and covalent bonding [86] have been used to immobilize biological molecules in conducting polymers.

Biomaterials need to be designed to physically support tissue growth and to elicit desired receptor-specific responses from specific cell types. One way to achieve these requirements is to incorporate biological molecules into synthetic materials. It has been shown that PPy is able to support the *in vitro* growth and differentiation of multiple cell types including neurons [58] and endothelial cells [87,88]. By choosing appropriate biologically active dopants, the properties of conducting polymers can be tailored for specific cell and tissue interactions.

For example, Collier et al. showed that hyaluronic acid, a hydrated glycosamino-glycan found in almost all extracellular tissues in the body, can be polymerized with PPy [89]. They assessed *in vitro* cellular response with PC 12 (pheochromocytoma) cells and *in vivo* response of the subcutaneous implantation of these materials in rats [89]. They suggested that the conductivity loss of a hyaluronic–PPy composite can be circumvented by using a bilayer approach. Garner et al. studied the incorporation of heparin, a potent anticoagulant known to promote endothelial cell growth, into PPy and assessed the resulting materials as substrates for endothelial cell growth. They suggested that when polymers were grown under conditions of equal total charge but at different current densities, those grown at higher current densities incorporated more heparin because the charge neutralization of PPy by heparin molecules that were already incorporated into PPy required some rearrangement of conformation that was kinetically limiting at higher rates of polymerization (i.e., higher current density) [87].

The incorporation of biomolecules into conducting polymers has been used for surface modification of implanted neural microelectrodes. Cui et al. synthesized PPy containing biomolecules including a silk-like protein with fibronectin fragments (SLPF) and a nonapeptide CDPGYIGSR onto microelectrode sites [4]. Incorporation of the nonapeptide CDPGYIGSR into PPy facilitated the growth of neuroblastoma cells. In addition, the higher surface area of the electrode sites facilitated charge transport, which is crucial for effective neural communication [4]. The stability of PPy–CDPGYIGSR coatings was tested in deionized water soaking experiments, where it was found that the peptides entrapped in PPy did not diffuse away within 7 weeks. More intensive *in vivo* assessment was subsequently performed on these materials, and it was found that significantly more neurofilament-positive staining was present on the peptide-coated electrodes than controls, indicating that the coatings had established strong connections with neurons. After 1 week in deionized water, 83% of the coated electrodes showed positive immunostaining for neurofilaments, while only 10% of the uncoated electrodes had this staining. At 2 weeks, 67% of the coated and 5% of the uncoated electrodes showed neurofilament staining [5].

Kim et al. incorporated nerve growth factor (NGF) as a counter-ion in the electrochemical deposition of PPy and PEDOT and evaluated the ability of NGF-incorporated PPy to elicit specific biological interactions with the neurons [90]. Impedance measurements at the biologically relevant frequency of 1 kHz revealed that the minimum impedance of the NGF-modified PPy film, 15 kΩ, was lower than the minimum impedance of the peptide-modified PPy film (100 kΩ). The PC-12 cells attached to the conductive PPy and PEDOT had extended neurites, indicating that the NGF in the polymer film remained biologically active (Figure 7.9). Thus, the incorporation of NGF can modify the biological interactions of the electrode without compromising the conductive properties of the polymeric film [90].

Richardson et al. incorporated the neurotrophin NT-3 into PPy films and showed increased neurite outgrowth from auditory neuron explants *in vitro* [168]. Electrical

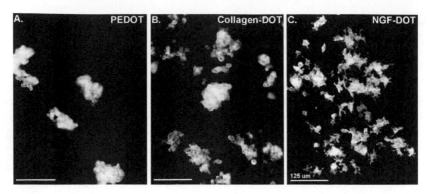

FIGURE 7.9 PC12 cells seeded on control PEDOT (A) as a control, on PEDOT–collagen (B), and PEDOT–NGF (C). Cells were incubated with each polymer substrate for 48 hours prior to imaging. Cells with neurites indicate that the NGF incorporated in the PEDOT film is not only bioactive but accessible to the cells. Because of their need for a substrate coated with collagen or a similar extracellular matrix adhesion molecule, the PC12 cells do not adhere well to PEDOT–PBS films (see [A]). However, PC12 cells adhere to PEDOT–protein composite films prepared with collagen, which also suggests that, like NGF, collagen is incorporated into the PEDOT film and is bioactive and accessible to the cells (compare [A and B]). (Adapted from Kim, D. H. et al., *Adv. Func. Mat.* 19(1), 79–86, 2007. With permission.)

stimulation was also shown to have a beneficial impact on neural regeneration. These films are being investigated as coatings on cochlear implants.

7.4.2 HYDROGELS

Hydrogels have been widely used in biomaterials applications because they can be highly swollen with water and have mechanical properties that are similar to those of living tissue. Alginate is a well-known polysaccharide obtained from brown algae that is widely used for drug delivery and tissue engineering. Alginate often serves as a delivery vehicle of cells for tissue engineering applications because of its biocompatibility, low toxicity, and simple gelation with divalent cations such as CaP^{2+}, MgP^{2+}, and SrP^{2+}. Alginates are a family of natural copolymers of β-D-mannuronic acid (M) and α-L-guluronic acid (G) [91]. A potential limitation of alginate gels in tissue engineering is the inherent resistance to protein adsorption and cellular attachment because of its hydrophilic character and lack of specific integrin-binding sequences such as the RGD tripeptide from fibronectin or the IKVAV and YIGSR sequences from laminin [92]. This nonfouling property may be useful for limiting the extent of glial scar formation on the neural electrodes after implantation.

Alginate hydrogels cross-linked with calcium sulfate ($CaSO_4$) have been used as cell delivery vehicles because of their slow process of gelation. Alginate beads are commonly produced by fast gelation with calcium chloride ($CaCl_2$) since it is more easily controlled than $CaSO_4$-induced gelation. However, this rapid gelation often results in a nonuniform structure [93]. To overcome the gradients of cross-link density and mechanical properties due to $CaCl_2$, $CaCO_3$ has been used to form more structurally uniform gels for use as scaffolding materials [94]. In the presence of other ions such as Na^+, ionically cross-linked alginate gels may lose mechanical strength or may swell via a process involving lost divalent ions into the surrounding medium and subsequent dissolution. The extent to which an alginate gel exhibits swelling depends on the ratio of Na^+ to Ca^{2+} and the composition of the alginate itself [95].

To improve integration with living tissue, hydrogels have been used to deliver and release growth factors and bioactive agents. Hydrogels containing biotrophic molecules such as vascular endothelial growth factor (VEGF) [96], basic fibroblast growth factor (bFGF) [97], epidermal growth factor (EGF) [98], and bone morphogenetic protein (BMP) [99] have been developed.

Kim et al. developed hydrogel coatings to improve the functionality of chronically implanted neural recording electrodes. An alginate hydrogel was used as the coating material on microelectrode arrays for better integration and mechanical buffering between the electrodes and CNS tissue (Figure 7.10). Kim et al. have confirmed that alginate hydrogels can be coated onto the surfaces of neural probes by a dipping process. The coating thickness can be varied from less than a micron to over 100 µm. During tissue insertion, problems can occur if the hydrogel coatings are too soft. In this case the coating can be sheared off the probe, remaining at the surface of the cortex. However, it has been found that it is possible to dehydrate the hydrogel and then allow the material to reabsorb water after insertion. The hydrogels have been shown to readily reswell after insertion into agar tissue phantoms (Figure 7.10) [9].

Acute studies of *in vivo* recordings from hydrogel-coated probes implanted in guinea pig brain were recently conducted by Kim et al. in collaboration with David J. Anderson and James Wiler at the Kresge Hearing Research Institute. Kim et al. examined the quality of recording as a function of the thickness using 200-msec noise bursts and found a significant reduction in the number of high-quality units recorded even for relatively thin (5 to 10 µm) films (Figure 7.11). Evidently, even these relatively thin hydrogels can push the neurons away from the electrode site. These results clearly demonstrated the importance of the local proximity of cells to the electrode surface and confirmed the need to establish intimate connections with the neurons to maintain signal quality [8]. Kim et al. also evaluated the use of PEDOT conducting polymer to maintain the functionality of hydrogel-coated

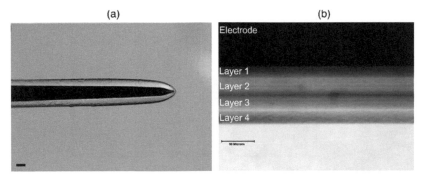

FIGURE 7.10 (a) Hydrogel coating deposited onto a microfabricated neural probe. (b) Multilayered hydrogel coating prepared by sequential dipping. Scale bar is 50 µm. (Adapted from Kim, D. H. et al., *Biomaterials* 27(15), 3031–3037, 2006. With permission.)

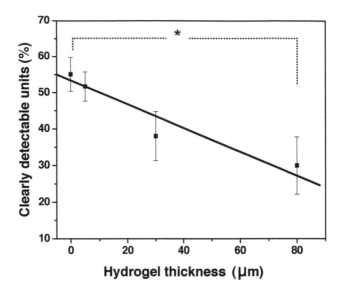

FIGURE 7.11 Decrease in fraction of clearly detectable units in guinea pig cortex for microfabricated neural probes coated with hydrogels of various thickness.

microelectrodes. PEDOT deposited on the electrode site restored the lost functionality of the electrodes caused by hydrogel coatings as shown by the number of clearly detectable units and the signal-to-noise ratio (Figure 7.12).

In designing materials that can facilitate charge transport through low-density structures, Kim et al. also found that conducting polymers can be grown in the hydrogels, resulting in open, extended networks that grow out from the electrode site into the hydrogel matrix. The conducting polymer precipitates onto the

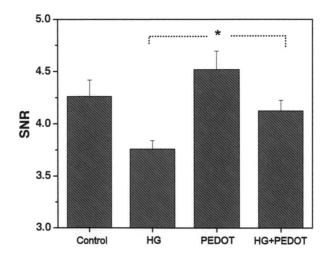

FIGURE 7.12 Average signal-to-noise ratios with various coatings including no coating, hydrogel (HG) coating, PEDOT deposition on the electrode sites, and PEDOT deposition on the electrode sites under the HG coatings (one-way ANOVA).

low-density hydrogel network as a scaffold. The result is an extremely intercon-nected, nanoporous, high-surface-area film (Figure 7.13). The impedances of these porous conducting polymer films are around three orders of magnitude less than the initial impedance of the metal electrode (Figure 7.14). Kim et al. also found that these films can be readily dehydrated and reswollen without apparent influence on the microstructure or electronic transport properties [9]. Furthermore, the conduct-ing polymers could still be readily grown through the hydrogel after disrupting the microstructure by freeze-drying. Impedance measurements at the biologically important frequency of 1 kHz showed that the minimum impedance of this poly-mer-modified hydrogel was 7 kΩ (Figure 7.14). This was much lower than the mini-mum impedance of PPy-coated electrodes (~100 kΩ) [9].

Antiinflammatory steroids such as dexamethasone (DEX) are of interest to reduce the tissue reaction. Kim et al. investigated the release of DEX from algi-nate hydrogel matrices and compared it with that from free nanoparticles (NPs) and NPs immobilized in the hydrogel matrix [100]. DEX-loaded poly(lactic-co-gly-colic acid) (PLGA) NPs with typical particle size ranging from 400 to 600 nm were prepared using a solvent evaporation technique. The *in vitro* release of DEX from NPs entrapped in the hydrogel showed that 90% of the drug was released over 2 weeks (Figure 7.15). The impedance of NP-loaded hydrogel coatings on microfab-ricated neural probes was measured and showed negligible increases over 3 weeks (Figure 7.16). *In vivo* impedance of chronically implanted electrodes loaded with

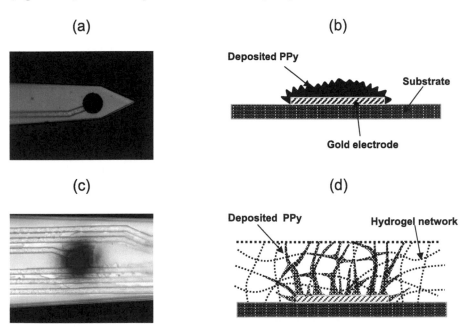

FIGURE 7.13 Optical microscope images of (a) film-like structure of conducting poly-mer PPy deposited on the electrode sites without a hydrogel scaffold. (b) Schematic of the film-like conducting polymer PPy. (c) Conducting polymer PPy grown through a hydrogel scaffold. (d) Schematic of the growth of the conducting polymer PPy through the hydrogel matrix.

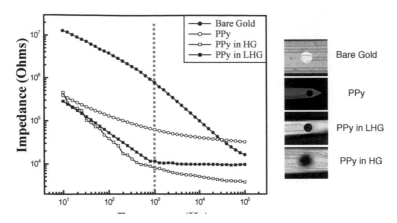

FIGURE 7.14   Impedance of PPy conducting polymers with different morphologies including bare gold, PPy, PPy in lyophilized hydrogel (LHG), and PPy in hydrogel (HG).

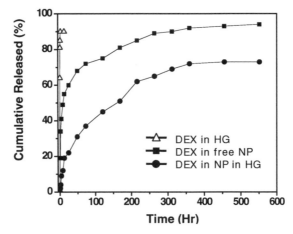

FIGURE 7.15   Amount of dexamethasone released from hydrogel without being incorporated into nanoparticles, from free nanoparticles, and from nanoparticles encapsulated into a hydrogel. (Adapted from Kim, D. H. et al., *Biomaterials* 27(15), 3031–3037, 2006. With permission.)

DEX was maintained at a relatively constant level, while that of control electrodes increased by three times about 2 weeks after implantation until it stabilized at approximately 3 MΩ [100] (Figure 7.16).

Tresco's group at the University of Utah has provided quantitative information about the nature and extent of cellular reactivity that occurs *in vivo* after the hydrogel-coated probes have been implanted into rat cortex for various periods of time (1 week, 2 weeks, and 12 weeks). They have confirmed the formation of a most proximal layer of microglia directly on the probe surface, followed by a layer of reactive astrocytes. There was also a reduction in the number of neurons in the corresponding area close to the probe. The microglia remained most active when the probes

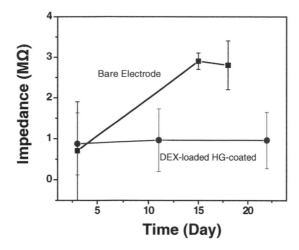

FIGURE 7.16   Impedance as a function of time for a dexamethasone-loaded, hydrogel-coated neural probe implanted into guinea pig cortex, compared with an uncoated control.

were mechanically tethered to the skull, supporting the hypothesis that mechanical strains from micromotion of the probe are involved in exacerbating the chronic inflammation [101]. Their most recent data also confirmed a substantial decrease in the amount of inflammation around hydrogel-coated probes.

7.4.3 NANOTUBES

There is currently considerable interest in the development of carbon-based nanotubes for a variety of applications. Several groups have investigated the use of carbon nanotubes for the surfaces of neural prosthetic devices. It has been found that polymer-modified carbon nanotubes can promote the differentiation and proliferation of NG108-15 and PC12 neuroblastoma cells [102,103]. Although there is some ability to functionalize the surfaces of these carbon nanotubes, the interactions of these stiff, rigid rods with living cells has not yet been determined. Another limitation is that most current methods of preparing nanotubes lead to a distribution in diameters and thus variations in physical properties (such as metallic or semiconducting). Methods to separate carbon nanotube mixtures based on their diameter should make it possible to better control their properties in the future [104]. Hybrid materials can also be produced. For example, it has recently been reported that nanocomposite coatings can be prepared by the electrochemical polymerization of conducting polymers such as PEDOT around carbon nanotubes [105].

Abidian et al. successfully established methods for the preparation of drug-loaded conducting polymer nanotubes [10]. The fabrication process involved the electrospinning of a biodegradable polymer (PLGA) into which a drug (dexamethasone) had been incorporated followed by electrochemical deposition of the conducting polymer around the drug-loaded electrospun biodegradable polymers (Figure 7.17). The diameters of the electrospun nanofibers ranged from 40 to 500 nm, with the majority between 100 and 200 nm. The wall thickness of the PEDOT

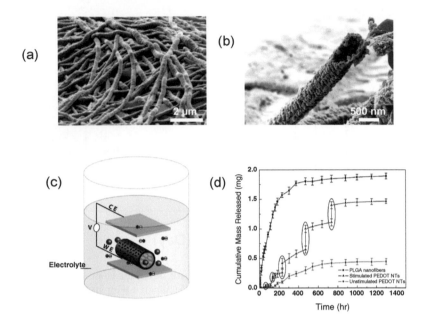

FIGURE 7.17 (a) SEM image of PEDOT nanotubes prepared by electrochemical polymerization around PLGA nanofibers. (b) Close-up view of the end of a single PEDOT nanotube after removal of the nanofiber template. (c) Schematic of the actuation of a nanotube under an externally applied potential. (d) Controlled spatial and temporal release of the anti-inflammatory agent dexamethasone from electrically stimulated PEDOT nanotubes. Stimulation leads to a local burst of released drugs a specific points in time (circled). (Adapted from Abidian, M. R. et al., *Adv. Mater.* 18(4), 405, 2006. With permission.) **(See color insert following page 110.)**

nanotubes varied from 50 to 100 nm, and the nanotube diameter ranged from 100 to 600 nm (Figure 7.17a,b). By controlling the polymerization time, tubular structures with thin walls (shorter deposition time) or thick walls (longer deposition time) could be reproducibly prepared. The initial impedance of the bare gold sites before surface modification was 800 ± 20 kΩ for acute probes (1250 μm^2). This value of impedance was decreased to a minimum of 8 ± 2 kΩ by formation of conducting polymer nanotubes on the electrode sites.

Abidian [10] also found that individual drugs and bioactive molecules could be precisely released at desired points in time by using electrical stimulation of nanotubes (Figure 7.17c,d). When a conducting polymer is exposed to an external electrical potential, ions will diffuse in or out of the material to balance the local electrostatic charge [106]. The extent of expansion or contraction depends on the number, size, and mobility of ions exchanged [107]. Electrochemical actuators using conducting polymers based on this principle have been developed by several investigators [108–110].

In collaboration with Joseph Corey, it has also been successfully demonstrated that aligned conducting polymer nanotubes can directionally guide the neurite outgrowth of DRG explants and PC12 cells (Abidian et al, unpublished).

7.4.4   HYBRID LIVE CELL-CONDUCTING POLYMER COATINGS FOR NEURAL ELECTRODES

The pursuit of a long-term, therapeutic brain–device interface continues to motivate advancements in the design and development of implantable neural prosthetic devices. However, unpredictable device performance associated with limited biocompatibility and poor tissue integration remains a barrier to successful testing and implementation [14,24,25,111–113]. To function properly once implanted, neural prosthetic devices rely on their ability to establish and maintain direct, functional communication with neurons in the surrounding tissue. This highlights the central importance of an intimate electrode–tissue interface and presents a challenge to bridging the biotic–abiotic interface by joining electronically and ionically conductive systems [115,116]. Intimate integration of electrodes with surrounding tissue will facilitate charge transfer between electrodes and target cells as well as increase biocompatibility.

Integration at the device–tissue interface can be increased through the use of bioactive or biomimetic materials that can physically and biochemically interact with surrounding tissue [117–119]. However, for use in implanted bioelectric devices these materials should also maintain or improve the electronic–ionic communication between the device electrodes and the tissue. Therefore, in recent years studies have focused on developing strategies to increase tissue integration, electrical sensitivity, and charge transfer capacity at the device–tissue interface through the use of inherently conductive polymers and conducting polymer–protein composites as low-electrical-impedance, bioactive coatings for microelectrodes on biomedical devices [59,62,74,89,103,120,121].

Numerous studies indicate that neural electrode functionality can be increased by modifying the surface of the electrode sites with low-impedance conductive polymer coatings with nanoscale roughness or porosity [4,6,7,11,79] and through the incorporation of cell adhesion peptides [4], proteins [31,90,122], or antiinflammatory drugs [100,123]. Together these studies suggest that the most benefit could be gained by multifunctional modifications of the electrode surface that have increased electrical activity, bioactivity, mechanical softness, and topological features on a similar scale to that of cells in tissues and cell surface and extracellular matrix structures. These ideas also evoke the possibility that the biofunctionality and biocompatibility of the electrode could be further increased by incorporating living cells or cellular components into an electrode coating to exploit unique cellular physiology and signal transduction capabilities.

Recent studies have found that living neural cells can be incorporated directly into a matrix of the conducting polymer, PEDOT, while still maintaining cell viability and signal transduction capabilities. This has resulted in the generation of functional hybrid PEDOT–neural cell electrode coatings as well as a method of using neural cells to "template" PEDOT films to create highly biomimetic conductive substrates with cell-shaped features that are also cell attracting. Electrical characterization of the conducting polymer matrix containing live neural cells suggested a relationship between the electrode and neural cells that is distinct from a more typical configuration used for electrically interfacing neurons in which neural cells are cultured on or near metal electrodes. Intimate interactions between the conducting polymer and the neuronal membrane were revealed as PEDOT covered delicate filopodia and neurites. This unique cell-conducting polymer–electrode interface may be an ideal candidate material for the development of a new generation of biosensors and "smart" bioelectrodes. The ability to polymerize PEDOT in the presence of living cells has been confirmed *in vitro* around living cells [165] as well as *in vivo* through living tissue [166].

In earlier studies described by the Wallace group, PPy was used to generate a novel biosensor composed of PPy doped with erythrocytes for detection of blood Rh-factor via the Rhesus factor antigens on the cell surface [124]. The erythrocyte-containing PPy bound three times as much antibody as unmodified PPy as detected by ELISA and resistometry. This study indicates that cell-conducting polymer matrices can perform sensitive and specific biosensing. However, Campbell et al. [124] found that the presence of erythrocytes within the PPy matrix did not alter the electrical properties of the PPy. This is likely because unlike neural cells, erythrocytes are nonadherent and nonelectrically active cells. The incorporation of electrically responsive, electrode-adherent cells into a conducting polymer matrix provides for an additional opportunity to utilize both the biochemical and electrochemical qualities of the incorporated cells for sensing purposes.

In preparation for experiments involving the polymerization of PEDOT in the presence of living cells, it was first determined whether PEDOT and its monomer EDOT were toxic to cells. Previous studies have shown that cells can be cultured for days to weeks on PEDOT and PPy with little or no toxicity [58,77]. However, the effect of the EDOT monomer on cell viability was not known. It was found that neural cells (SH-SY5Y neuroblastoma-derived cell line) could be exposed to as much as 0.01 M EDOT and 0.02 M PSS (polyanionic dopant in monomer solution) for as long as 72 hours while maintaining 80% cell viability (Figure 7.18). In our studies,

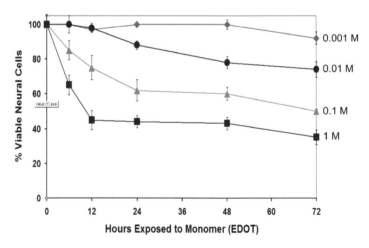

FIGURE 7.18 Working concentrations of EDOT monomer are not cytotoxic. MTT cytotoxicity assay for exposure of SH-SY5Y neural cells to increasing concentrations of EDOT in monomer solution (all with 0.02 M PSS) for 0 to 72 hours. (Adapted from Richardson-Burns, S. M. et al., *Biomaterials*, 28(8), 1539–1552, 2007. With permission.)

cells were typically exposed to EDOT for less than 10 minutes, so the cytotoxicity was negligible.

PEDOT was electrochemically polymerized directly in the presence of live neural cells cultured on electrodes (Au, Au/Pd, or ITO) using 0.5 to 1 uA/mm² galvanostatic current from a monomer solution containing 0.01 M EDOT and 0.02 M PSS in phosphate buffered saline (PBS) (Figure 7.19a,b). This resulted in the formation of PEDOT on the electrode, surrounding and embedding the cells (Figure 7.19c,d,e). The morphology and topology of the PEDOT polymerized around the neural cells was assessed using optical microscopy and scanning electron microscopy (SEM). After deposition, PEDOT appeared as a dark, opaque substance around the cells. The cells themselves and their nuclei remained intact throughout and following polymerization (Figures 7.19c and 7.19e, respectively). Interestingly, PEDOT deposition was prohibited in areas where cells were evidently strongly adhered to the substrate (Figure 7.19c). Using SEM, it was found that the PEDOT on the electrode and around the cells displays the fuzzy, nodular surface topology that is typical of PEDOT (Figure 7.20A). The polymer also appeared to wrap around the exterior of the cells and their extensions (Figure 7.20B,C).

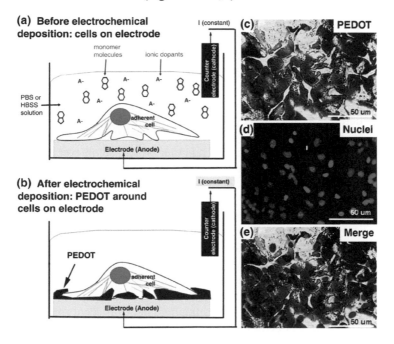

FIGURE 7.19   The conducting polymer PEDOT can be electrochemically polymerized in the presence of living cells. (a) Diagram representing the electrochemical deposition cell and the neural cell monolayer cultured on the surface of the metal electrode prior to polymerization. (b) Diagram representing PEDOT polymerized around living cells. (c) PEDOT (dark substance) polymerized in the presence of a monolayer of SY5Y neural cells cultured on electrode. (d) Nuclei of SY5Y cells stained with Hoechst 33342 (blue florescence). (e) Merged image showing nuclei of cells around which PEDOT is polymerized. (Adapted from Richardson-Burns, S. M. et al., *Biomaterials*, 28(8), 1539–1552, 2007. With permission.) **(See color insert.)**

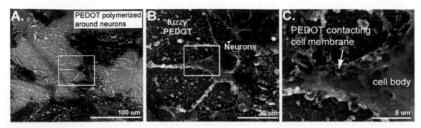

FIGURE 7.20   SEM images of PEDOT polymerized around neural cells. (A) PEDOT polymerized in the presence of mouse primary dissociated cortical cultures (MCC). PEDOT (rough, nodular texture) covers the electrode surface as well as some cellular processes. (B) Higher-magnification image of area in box from (A) reveals that PEDOT uses the cell membrane as a scaffold for polymerization. (C) Higher-magnification image of area in box from (B) further reveals the intimate electrode–cell interface resulting from polymerizing PEDOT around living neurons. The PEDOT that contacts and partially covers the cell membrane is pseudocolored to indicate the intimate polymer–membrane contact that is achieved via electrochemical polymerization. (Adapted from Richardson-Burns, S. M. et al., *J. Neural Eng.* 4(2), L6–L13, 2007. With permission.)

Neural cells partially embedded in PEDOT maintained their viability for almost 1 week, suggesting that the PEDOT matrix was not a significant barrier to cell nutrient transport. However PEDOT-surrounded neurons eventually began to die by apoptosis that can be triggered as long as 24 to 72 hours after the initial insult [125,126]. One of the more dramatic findings in cells surrounded by PEDOT was substantial disruption of the cytoskeleton, specifically a loss of F-actin stress fibers that can be detected as early as 2 hours after polymerization and is complete by 24 hours after polymerization (Figure 7.21A–H). These fibers are physically and biochemically associated with integrins and other protein complexes at focal adhesions that are the primary mediators of cell surface interactions with the extracellular matrix and neighboring cells [127]. This observation provides insight about the molecular-level interactions at the plasma membrane–PEDOT interface and suggests that the presence of the polymer so near the membrane may disrupt integrin signaling and focal adhesion maintenance [128,129]. Furthermore, loss of F-actin stress fibers is indirect evidence of the activation of biochemical signaling pathways involving focal adhesion kinase (FAK), jun-N terminal kinase (JNK), and src (tyrosine kinase oncogene), each of which has been implicated in apoptosis [130–132]. This may explain why cells embedded in PEDOT undergo apoptosis. It also gives us molecular targets for future studies to attempt to interrupt focal adhesion disruption, block development of cytoskeletal abnormalities, and rescue cells from death. In addition, studies of neuronal apoptosis have indicated that alterations in actin cytoskeletal morphology can be associated with oxidative stress in neurons [133]. The PEDOT electropolymerization process may involve production of free radicals at or near the surface of the neurons on the electrode, so future studies will also explore whether oxidative stress plays a role in disruption of the actin cytoskeleton in neurons embedded in PEDOT.

PEDOT was polymerized in the presence of live neural cells to generate conductive polymer substrates with biomimetic topology consisting of cell-shaped holes and imprints on the same scale as cell surface features. Following polymerization

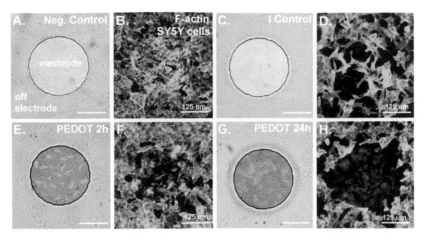

FIGURE 7.21 The F-actin cytoskeleton is disrupted in cells embedded in the PEDOT matrix. (A) Bright-field image of SY5Y cells cultured on electrode (negative control; no PEDOT, not exposed to current). (B) SY5Y cells cultured on electrode (negative control; no PEDOT, not exposed to current) stained for F-actin with Phalloidin-Oregon Green (green fluorescence). (C, D) Cells exposed to current only (no PEDOT), fixed and stained 24 hours after current exposure, show some morphological alterations but F-actin staining remains robust. (E, F) Cells 2 hours after PEDOT polymerization begin to show abnormal cytoskeletal staining and loss of F-actin stress fibers. (G, H) Cells 24 hours after PEDOT polymerization show a nearly complete loss of F-actin stress fibers and a decrease in the intensity of F-actin staining overall. (Adapted from Richardson-Burns, S. M. et al., *Biomaterials*, 28(8), 1539–1552, 2007. With permission.)

of PEDOT around the neurons, the cells and cell material were removed from the PEDOT matrix using enzymatic and mechanical disruption. This resulted in a neural cell-templated, fuzzy PEDOT material with a combination of nanometer and micrometer scale features (Figure 7.22A,B,C). The neural cell-templated polymer topography included neuron-shaped holes (Figure 7.22A) and tunnels, crevasses, and caves (Figure 7.22B,C) resulting from conductive polymer molded around cell bodies and extended neurites. Through use of this method, evidence was found of intimate contact at the interface between the PEDOT matrix and plasma membrane of the cells as exemplified in Figure 7.22, in which the PEDOT (dark substance) revealed nanometer-scale tendrils at the leading edge of a neurite (Figure 7.21D,E,F).

It was hypothesized that the biomimetic surface of the cell-templated PEDOT would be attractive to cells because of its nanometer scale fuzziness and the unique cell-shaped holes and imprints. Therefore it was tested whether new cells seeded on top of the cell-templated PEDOT would show evidence of repopulating the cell-shaped holes or of increased adhesion to the cell-templated surface. It was found that SY5Y cells cultured on the neuron-templated PEDOT substrate showed a preference for adhering to the cell-templated zones over the regions of untemplated PEDOT (Figure 7.22A,B,C). A subset of cells did seem to repopulate the cell-shaped holes of the film (Figure 7.22A,B,C), however these cells did not settle down into the exact position of the original cells used for templating. These findings suggest that when

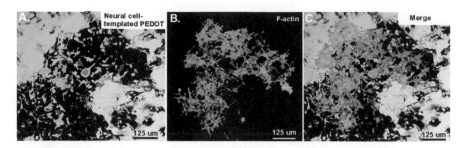

FIGURE 7.22 Neuron-templated PEDOT films are cell-attracting, conductive substrates. (A) Optical image of neural cell-templated PEDOT film on electrode shows numerous cell-shaped holes and neurite-templated channels left behind following removal of cells from the PEDOT matrix (dark substance). (B) A new monolayer of SY5Y cells are cultured on the neuron-templated PEDOT film. These cells are stained with Phalloidin-Oregon Green (green fluorescence; F-actin) to visualize cell morphology and position of cellular processes. (C) Merged image suggests that the new cells show preferential adhesion to neuron-templated PEDOT and some cells repopulate the cell-shaped holes in the polymer film. (Adapted from Richardson-Burns, S. M. et al., *Biomaterials*, 28(8), 1539–1552, 2007. With permission.)

implanted in tissue, this cell-templated polymer surface may encourage cells in the host tissue to adhere near or within the cell-shaped holes and send processes into the tunnels and crevasses. This would provide for very intimate contact between cells and the conductive polymer, making possible continuous electrical contact between the electrode and the tissue. The use of different methods for removing cells results in variation in the amount of cell material that remains associated with the PEDOT matrix. This provides an opportunity for spatially localized biochemical control of interactions between target cells and the electrode at the cellular- and subcellular-length scale. When coupled with the mechanical control provided by the cytomimetic topology, tailoring of the biochemistry of the cell-templated surface could make possible precise manipulation and tracking of neurite guidance, growth, and signal transduction.

7.4.5 SURFACE CHEMICAL CHARACTERIZATION BY X-RAY PHOTOELECTRON SPECTROSCOPY (XPS)

Recently, an XPS study comparing Baytron P, a form of commercially available PEDOT–PSS, and electrochemically deposited PEDOT–PSS was completed. The following sections gives a background into previous XPS work completed on PEDOT with counter-ions, followed by studies on PEDOT degradation mechanisms.

7.4.5.1 PEDOT and Counter-Ions

The majority of XPS characterization has been completed on the commercially available PEDOT–PSS, Baytron P, because of its use in organic electronic devices [134]. The characteristic regions normally used for PEDOT analysis are the carbon (C 1s), oxygen (O 1s), and sulfur (S 2p) regions. These initial studies focused on the effect of different dopants on the PEDOT binding energy in an effort to deduce how the counter-ion was binding with PEDOT.

Initial XPS characterization was carried out on chemically polymerized PEDOT via iron(III) tris-p-toluene sulfonate. PEDOT was found to have peaks at 289.8 ± 0.2 eV in the C 1s region, 538.4 ± 0.2 eV in the O 1s region, and 168.2 eV in the S 2p region [135]. The PEDOT was then doped with the large polymeric anion PSS- or the small anion tosylate (p-methyl benzyl sulfonate) (TsO-). The normal peak positions for poly (sodium 4-styrenesulphonate), PSS-Na$^+$, are 285.0 eV (aliphatic carbon), 284.68 eV (aromatic carbon), 285.16 eV (C-S), 531.72 eV (O 1s in SO$_3$-), 168.23 eV (S 2p$_{3/2}$ in SO$_3$-), and 1071.76 eV (Na$^+$) [136].

Xing et al. found that the use of anionic counter-ions resulted in the oxidation of PEDOT, thus broadening the C 1s peaks. The C 1s for PEDOT–PSS- was found at 289.4 eV, while the peak for the PEDOT–TsO- was at 288.8eV [135]. The addition of PSS- resulted in a second peak at 536.4 ± 0.2 eV and a slight shift in the PEDOT peak (538.4 ± 0.2 eV) in the O 1s region, while the use of TsO- resulted in a second peak at 535.8 eV along with the PEDOT peak at 538.2 eV in the O 1s range. Both PSS- and TsO- exhibited two peaks at 172 eV and 168.2 eV in the S 2p range [135]. Additional tosylate work has also been performed by Kim et al. [137].

Greczynski et al. continued to study the effect of counter-ions on PEDOT by studying the dopants poly(4-styrenesulfonic acid) (PSSH) and PSS-Na$^+$ [138,139]. Commercial PEDOT–PSS was found to have S 2p peaks at 164.5 and 165.6 eV (spin-orbit coupling splitting of the PEDOT sulfur) and a peak at 169 eV (PSS). Peak deconvolution analysis showed the S 2p$_{3/2}$ of PSSH to be at 168.8 eV and the S 2p$_{3/2}$ of PSS-Na$^+$ at 168.4eV [138,139]. Greczynski et al. found that the S 2p spectrum is complicated by the presence of PEDOT. This manifests itself in an asymmetric tail at higher binding energies, a general shift to higher binding energies, and broad binding energy distributions, resulting from a positive charge delocalized over multiple and adjacent rings. Analysis of the O 1s peak found that the peak was composed of multiple peaks at 532.4 eV (oxygen double bonded to sulfur in PSSH), 533.5 eV (hydroxyl-oxygen atoms), 531.9 eV (PSS-Na$^+$), and 533.7 eV (oxygen atoms in the dioxyethylene bridge of PEDOT) [138,139].

Zotti et al. studied the doping structure relationship of electrochemically polymerized PEDOT with p-toluenesulfonic acid (TosH), sodium toluenesulfonate (TosNa), PSSH, and PSS-Na$^+$ [140]. This study, which also compared the chemically polymerized PEDOT–PSS to the electrochemically polymerized PEDOT–PSS, found that the electrochemically polymerized PEDOT/PSS contained more PEDOT relative to PSS than the chemically polymerized version through comparison of the PSS to thiophene ring (PEDOT) ratios. Evidently, electrochemically polymerized PEDOT–PSS was more compact, allowing for a smaller distance between chains, resulting in a reduction of the charge hopping distance. It is also probable that a single PSS polyanion connects multiple PEDOT chains, ultimately leading to higher conductivity [140].

An S 2p region comparison of the counter-ions between the two electrochemically polymerized PEDOT–PSSH and PEDOT–TosH samples yielded peaks at 167.8 eV (SO$_3$-H$^+$) and 166.8 eV (SO$_3$-PEDOT$^+$) for the PEDOT–PSSH and at 166.8eV (SO$_3$-PEDOT$^+$) for the PEDOT–TosH. From the integrated peak intensities, the relative quantities of the counter-ions were deduced. The PSS counter-ion was found to be present in a greater amount than the TosH counter-ion because only half of the PSS can be used to neutralize the PEDOT charge, whereas the amount of TosH utilized is only that which is needed to neutralize the PEDOT [140,141]. The greater amount of PSS results in a decrease in conductivity due to the increase in the average charge hopping distance [140].

Based on the PSS to thiophene ring ratio, PEDOT–PSSH had a smaller amount of sulfonate in the film than the PEDOT–PSS⁻Na⁺ samples, and a greater quantity of PEDOT within the PEDOT–PSSH samples was found than was present in the PEDOT–PSS-Na⁺. PSSH was therefore found to dope the PEDOT better than the PSS-Na⁺ [140].

XPS has also been utilized to deduce the chemical binding energies of PEDOT after surface treatments, such as an acid or heat treatment, and under different solvent conditions. The Greczynski et al. study described the effects of hydrochloric acid (HCl) treatment and thermal treatments on the commercial PEDOT–PSS [138,139]. The effects of using different solvent solutions containing sorbitol, N-methylpyrrolidone (NMP), and isopropanol when heated were studied by Jönsson et al. [167].

7.4.5.2   PEDOT Degradation

PEDOT degradation studies using XPS have also been conducted. These studies focus on the atmospheric [141], UV-light [141,142], and electron bombardment degradation mechanisms [141,143].

Atmospheric effects on PEDOT and PSS, specifically the appearance of nitrogen, were studied by Crispin et al. [141]. The appearance of nitrogen within PEDOT–PSS films is thought to be a result of atmospheric ammonia molecules (NH_3) reacting with water and the sulfonic acid group of PSS to form a hydroxide, which further reacts to form ammonium salts. The formation of ammonium salts induces desulfonation and thus aging of the PSS, which has been known to occur with exposure to light and heat [141].

Marciniak et al. [142] explored ultraviolet (UV) light degradation on EDOT, EDOT–SO_2, PEDOT–PF_6, and PEDOT–$C_{14}H_{29}$. After UV exposure, the presence of SO_2 peaks indicated that photo-oxidation had occurred, resulting in shorter conjugation lengths that ultimately reduced conductivity. Chain scission was also found to occur. Marciniak et al. suggest that the degradation pathway of PEDOT is similar to the pathway in polythiophenes, in which energy transfer to the oxygen follows the $\pi-\pi^*$ transition, and then the oxygen finally reacts with the conjugated chain [142].

Electron bombardment studies were carried out on films of commercial PEDOT–PSS, PSSH, and EDOT to study the degradation effects on the polymer's structure [141,143]. The result of this study showed that the energy used (3 eV [current density: 1 $\mu A/cm^2$] for 67 h), which is less than typical energies used in organic light-emitting devices, was enough to trigger degradation of PEDOT. The degradation of the PEDOT thus involves not only the deterioration of the charge transport mechanism resulting in lower conductivities, but also the formation of free oxygen and sulfur atoms that in turn could cause breakdowns in other layers of the device because of their highly reactive nature [143].

In addition to deducing information on various counter-ion bonding, surface treatments, and degradation pathways, XPS has also been used to verify the presence of additives within PEDOT or PEDOT derivatives, such as PEDOT–PSS with poly(ethylene glycol) [144], adenosine triphosphate (ATP) [145], gold NPs [146], and PEDOT-coated latex spheres [147]. Other studies have focused on PEDOT binding with substrate materials, such as aluminum [148] and indium tin oxide [52,149].

7.4.5.3   Current Work

Figure 7.24 shows a schematic of the structure of an electrochemically polymerized PEDOT–PSS film, as well as a spun-cast film from a chemically polymerized PEDOT–PSS suspension. The PEDOT chains are relatively rigid, whereas the PSS is more flexible. The anionic PSS chains are expected to aggregate around the PEDOT. In the electrochemically polymerized film, the reaction proceeds by charge transport from the metal electrode out toward the surrounding solution. Presumably, the closest chains connect directly to the metal, with bonds that have a certain degree of charge transfer depending on the differences in electronegativity between the metal substrate and the PEDOT moieties. As the film thickens and densifies, occasional branching points allow for a rough surface texture as indicated. However since charge must be transferred from the metal substrate to the surrounding solution for the conducting polymer film to electrochemically deposit, efficient electrical pathways need to maintained.

For the PEDOT/PSS film prepared by chemical oxidation, the polymer chains are first formed in solution and are surrounded by PSS to help keep the molecules suspended. The film is formed by evaporation onto the substrate. In this case, close interactions between chains are less likely to form, and there may be a preferred orientation of the chain backbone in the plane of the film. There is also substantially more PSS in the film, as confirmed by XPS studies (Figure 7.23). Because of these differences in structure, electrochemically polymerized films typically show much better electrical properties than solution-deposited films of PEDOT–PSS and are therefore generally preferred for biomedical device applications.

7.4.6   Biological Conjugated Polymers: Melanins

Although PEDOT is not known to be found in its native state in living systems, examples exist of conjugated polymers with quite similar chemistries that are common in nature. Important examples are the melanins, which function as the dark, light-absorbing molecules that color hair and skin. Melanin derivatives are also found in certain tissues such as the eyes and ears and in the substania nigra of the human brain. It is known that reduced levels of melanin production in the brain correlate with certain disease states. For example, the loss of neuromelanin in the brain is associated with Parkinson's disease [150]. Although natural and synthetic melanins have been studied relatively extensively in the literature, their detailed biological function and potential use in biomedical devices is still not well established [151–153].

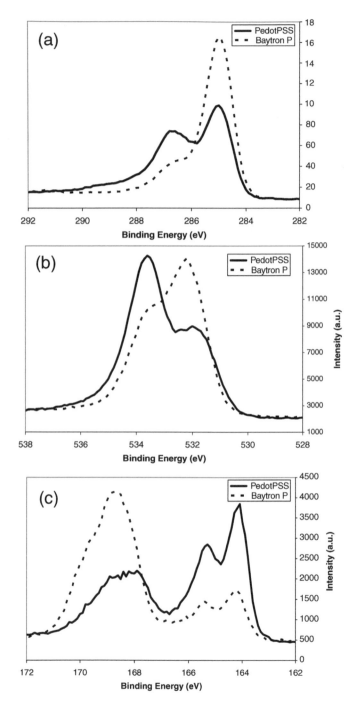

FIGURE 7.23 XPS spectra of (a) carbon (C 1s), (b) oxygen (O 1s), and (c) sulfur core levels (S s2p) for electrochemically polymerized PEDOT–PSS (solid) and Baytron P (dashed). The changes in relative peak intensity indicate that there is more PSS relative to PEDOT in the Baytron P sample.

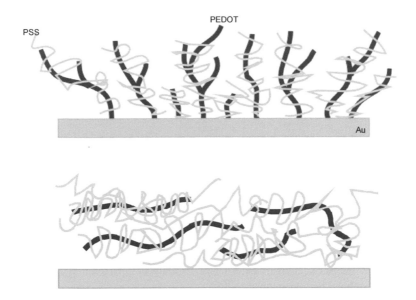

FIGURE 7.24 Schematic diagram comparing the microstructure of electrochemically polymerized PEDOT (top) with solution-cast films of commercially chemically polymerized PEDOT–PSS suspension (Baytron P). Since the electrochemically polymerized PEDOT requires electronic charge to grow, there must be electrical pathways to the metal substrate. It is expected that in these films the molecules are preferentially standing perpendicular to the surface with occasional branches, facilitating transport to the metal substrate. Solution cast films do not necessarily maintain good electrical connections between the PEDOT backbones, and XPS studies confirm that they generally have more of the flexible PSS counter-ion. It is also expected that the relatively rigid PEDOT backbones will tend to become aligned parallel to the sample surface.

Eumelanin and pheomelanin, the two main forms of melanin, are both composed of aromatic, fused bicyclic repeating units. Eumelanin contains both 5,6-dihydroxy-indole (DHI) and 5,6-dihydroxyindole-2-carboxylic acid (DHICA) repeating units. As shown in Figure 7.25, these monomers have two oxygen atoms directly attached to a conjugated molecular backbone. This chemistry is also found in EDOT. It is therefore reasonable to anticipate that melanin may have electrical activity similar to that of PEDOT. Also, since eumelanin is an all natural, totally biologically derived polymer, it may have advantages for interfacing with living tissue. However, it may also prove to be less chemically stable over the long term.

Zielinski and Pande reported the first electrochemical synthesis of eumelanin in 1990 [154]. Horak and Weeks synthesized melanin from DHI in phosphate buffer in 1993 [155]. Since then eumelanin has been synthesized using 3,4-dihydroxy-L-phe-nylalanine (L-dopa) (Figure 7.26) [151–153] and DHI monomers [156,157]. L-dopa is the natural precursor for eumelanin but does not contain an indde unit. Cycliza-tion of the molecule occurs during polymerization to form both types of eumelanin repeating units shown in Figure 7.25. Compared to L-dopa, the structure of DHI has more of a resemblance to the repeating unit structure of eumelanin, but DHI is

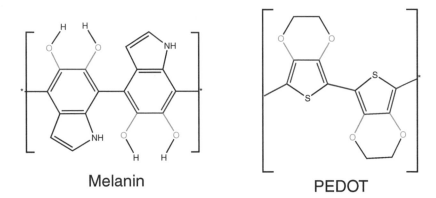

Melanin   PEDOT

FIGURE 7.25 Chemical structures of melanin and PEDOT. Note that both molecules share a conjugated molecular backbone, and they have oxygen atoms pendent to the backbone that in PEDOT are known to donate electrical charge and improve the electrical properties and chemical stability. Both polymers are efficient absorbers of light, leading to dark brown (melanin) or dark blue–black appearances (PEDOT).

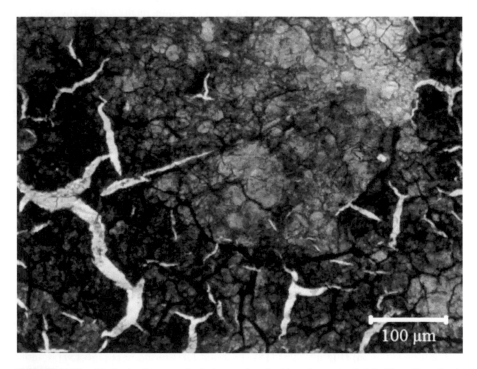

FIGURE 7.26 Optical micrograph of electrochemically polymerized thin film of synthetic melanin.

unstable in air and requires an inert atmosphere during all synthetic steps. Although DHICA is also a repeating unit in natural eumelanin, the direct electrochemical polymerization of DHICA has not yet been reported.

Studies of the electrical and optical properties of eumelanin imply that it is a semiconducting polymer [158,159]. Free-standing films of electropolymerized dopa-melanin were shown to have a conductivity of 1.4×10^{-6} S/cm [153], while dopa-melanin polymerized oxidatively has a conductivity of 6.4×10^{-8} S/cm [160]. Many synthetic eumelanins have also demonstrated photoconductivity [153,158]. There are still interesting open questions about the chemical structures that are formed during eumelanin polymerization. It appears that eumelanin consists of oligomers that contain many chemically distinct species. These species could include DHI, DHICA, and the oxidized forms of these molecules. Also, it seems that the monomer units bond to each other through many different types of atoms and that there is a general lack of organization within the oligomers [161,162]. Since the type of microstructure that forms should affect the electrical properties of the eumelanin film, controlling the structure to create long, completely conjugated polymer chains might result in a more electrically useful material.

Although it has not yet been studied in much detail, there have been conflicting studies concerning the effects of synthetic eumelanin on living cells. Ostergren et al. reported the ability to load PC12 cells with synthetic dopa-melanin without affecting the viability of the cells [163]. However, Li et al. showed that dopa-melanin-loaded SK-N-SH cells demonstrate increased cellular stress and that death occurs via apoptotic mechanisms [164]. The increased cell death is believed to be at least partially due to the formation of hydroxyl radicals. If this is the case, it may be possible to modify the hydroxyl groups on eumelanin to eliminate the formation of free radicals.

7.5 CONCLUSIONS

The electroactive, biomimetic conducting polymer and hybrid conductive polymer–live cell electrode coatings described here represent novel strategies for addressing shortcomings at the electrode–tissue interface. This technology should help facilitate the establishment of long-term, bidirectional communication between host cells and implanted microelectrode-based biomedical devices that is critical for realization of functional body–machine interfaces. These studies further elucidate the factors that are important in the design and development of novel electroactive biomaterials intended for direct, functional contact with living electrically active tissues including the CNS, PNS (peripheral nervous system), sensory organs, and heart.

ACKNOWLEDGMENTS

We gratefully acknowledge the financial support of our research from the National Institutes of Health National Institute of Neurological Disorders and Stroke (NIH-NINDS-N01-NS-1-2338), the National Science Foundation (DMR-0518079), the University of Michigan College of Engineering Gap funding award, and the U.S.

Army Multidisciplinary Research Initiative (MURI) program on Bio-Integrating Structural and Neural Prosthetic Materials. We have benefited from productive collaborations with Joseph Corey, Eva Feldman, Yoash Raphael, Bryan Pfingst, Wayne Aldridge, Frank Pelosi, Daryl Kipke, Paul Cederna, and Steve Goldstein at the University of Michigan. We have also worked in collaboration with Patrick Tresco at Utah, Greg Kovacs at Stanford, Jim Weiland at USC, Richard Normann at Utah, and Richard Stein at the University of Alberta. A variety of undergraduate research assistants provided laboratory assistance, including Wynn Koehler (MSE), Matt Meier (BME), Matthew Lapsley (MSE), Tani Kahlon (MSE), Mark Ferrall (LS&A), Michelle Leach (BME), Clair Harris (LS&A), Eric Tannebaum (LS&A), Deepa Rengaraj (MSE), Amber Brannan (Rose-Hulman Institute of Technology), Catherine Burk (BME), Kyle Roebuck (BME), Jingga Morry (BME), Brian Foster (EECS), Elizabeth Flak (MSE), Kate Gallup (MSE), Sejal Tailor (MSE), Beneque Cousin (MSE), Daniel Margul (BME), Olivia Kao (BME), and Grace Hu (MSE). Many of these undergraduate students were supported by the University of Michigan Undergraduate Research Opportunity Program (UROP). We also recognize contributions from Jayne Choi, Peter Keshtkar, and Max Betzig from Greenhills High School in Ann Arbor, who participated in a summer research outreach program coordinated by Rachel Goldman of the MSE Department. We also recognize the efforts of Rickard Axelsson, who visited our laboratory from Olle Inganas' group in Sweden. Several invention disclosures related to this research activity have been filed with the University of Michigan Office of Technology Transfer, and patents are pending at the U.S. Patent and Trademark Office. Various aspects of this work are under active consideration for potential commercialization. David C. Martin, Sarah Richardson-Burns, and Jeffrey L. Hendricks are cofounders of Biotectix LLC, a recently formed start-up company that has obtained exclusive options to license aspects of this work from the University of Michigan.

REFERENCES

1. Rousche, P. J. and Normann, R. A. Chronic recording capability of the Utah Intracortical Electrode Array in cat sensory cortex. *Journal of Neuroscience Methods* 82, 1–15, 1998.
2. Nicolelis, M. A. et al. Chronic, multisite, multielectrode recordings in macaque monkeys. *Proceedings of the National Academy of Sciences USA* 100, 11041–11046, 2003.
3. Buchko, C. J., Slattery, M. J., Kozloff, K. M., and Martin, D. C. Mechanical properties of biocompatible protein polymer thin films. *Journal of Materials Research* 15, 231–242, 2000.
4. Cui, X. et al. Surface modification of neural recording electrodes with conducting polymer/biomolecule blends. *Journal of Biomedical Materials Research* 56, 261–272, 2001.
5. Cui, X., Wiler, J., Dzaman, M., Altschuler, R., and Martin, D. C. *In-vivo* studies of polypyrrole/peptide coated neural probes. *Biomaterials* 24, 777–787, 2003.
6. Yang, J. and Martin, D. C. Microporous conducting polymers on neural prosthetic devices. II. Physical characterization. *Sensors and Actuators A: Physical* 113, 204–211, 2004.

7. Yang, J. and Martin, D. C. Microporous conducting polymers on neural prosthetic devices. I. Electrochemical deposition. *Sensors and Actuators B: Chemical* 101, 133–142, 2004.

8. Kim, D. Surface modification of neural prosthetic devices by functional polymers incorporating neurotrophic and pharmacological agents. *Biomedical Engineering* 148, 2005.

9. Kim, D., Abidian, M., and Martin, D. C. Conducting polymers grown in hydrogel scaffolds coated on neural prosthetic devices. *Journal of Biomedical Materials Research* 71A, 577–585, 2004.

10. Abidian, M. R., Kim, D.-H., and Martin, D. C. Conducting polymer nanotubes for controlled drug release. *Advanced Materials* 18, 405–409, 2006.

11. Ludwig, K. A., Uram, J. D., Yang, J., Martin, D. C., and Kipke, D. R. Chronic neural recordings using silicon microelectrode arrays electrochemically deposited with a poly(3,4-ethylenedioxythiophene) (PEDOT) film. *Journal of Neural Engineering* 3, 59–70, 2006.

12. Heidushka, P., Romann, I., Stieglitz, T., and Thanos, S. Perforated microelectrode arrays implanted in the regenerating adult central nervous system. *Experimental Neurology* 171, 1–10, 2001.

13. Bai, Q., Wise, K. D., and Anderson, D. J. A high-yield microassembly structure for three-dimensional microelectrode arrays. *IEEE Transactions on Biomedical Engineering* 47, 281–289, 2000.

14. Williams, J. C., Rennaker, R. L., and Kipke, D. R. Stability of chronic multichannel neural recordings: Implications for a long-term neural interface. *Neurocomputing* 26–27, 1069–1076, 1999.

15. Porada, I., Bondar, I., Spatz, W. B., and Krüger, J. Rabbit and monkey visual cortex: more than a year of recording with up to 64 microelectrodes. *Journal of Neuroscience Methods* 95, 13–28, 2000.

16. Edgerton, B. J., House, W. F., and Hitselberger, W. Hearing by cochlear nucleus stimulation in humans. *Annals of Otology, Rhinology, and Laryngology Supplement* 91, 117–124, 1982.

17. Benabid, A. L. et al. Long-term suppression of tremor by chronic stimulation of the ventral intermediate thalamic nucleus. *Lancet* 337, 403–406, 1991.

18. Bell, T. E., Wise, K. D., and Anderson, D. J. A flexible micromachined electrode array for a cochlear prosthesis. *Sensors and Actuators A–Physical* 66, 63–69, 1998.

19. Miller, K. and Chinzei, K. Modelling of brain tissue mechanical properties: Bi-phasic versus single-phase approach. *Proceedings of the 3rd International Symposium on Computer Methods in Biomechanical and Biomedical Engineering*, 1997.

20. Miller, K. and Chinzei, K. Mechanical properties of brain tissue in tension. *Journal of Biomechanics* 35, 483–390, 2002.

21. Mensinger, A. F. et al. Chronic recording of regenerating VIIIth nerve axons with a sieve electrode. *Journal of Neurophysiology* 83, 611–615, 2000.

22. Rousche, P. J. et al. Flexible polyimide-based intracortical electrode arrays with bioactive capability. *IEEE Transactions on Biomedical Engineering* 48, 361–371, 2001.

23. Seymour, J. and Kipke, D. R. Open-Architecture Neural Probes Reduce Cellular Encapsulation. Paper 0926-CC02-04, Materials Research Society, 2006.

24. Szarowski, D. H. et al. Brain responses to micro-machined silicon devices. *Brain Research* 983, 23–35, 2003.

25. Edell, D. J., Toi, V. V., McNeil, V. M., and Clark, L. D. Factors influencing the biocompatibility of insertable silicon microshafts in cerebral cortex. *IEEE Transactions on Biomedical Engineering* 39, 635–643, 1992.

26. Wise, K. D., Anderson, D. J., Hetke, J. F., Kipke, D. R. and Najafi, K. Wireless implant-able microsystems: High-density electronic interfaces to the nervous system. *Proceedings of the IEEE* 92, 76–97, 2004.
27. Landis, D. M. D. The early reactions of nonneural cells to brain injury. *Annual Review of Neuroscience* 17, 133–151, 1994.
28. White, R. L., Roberts, L. A., Cotter, N. E., and Kwon, O. H. Thin-film electrode fabrication techniques. *Annals of the New York Academy of Sciences* 405, 183–190, 1983.
29. Moxon, K. A., Leiser, S. C., Gerhardt, G. A., Barbee, K. A., and Chapin, J. K. Ceramic-based multisite electrode arrays for chronic single-neuron recording. *IEEE Transactions on Biomedical Engineering* 51, 647–656, 2004.
30. Yuen, T. G. and Agnew, W. F. Histological evaluation of polyesterimide-insulated gold wires in brain. *Biomaterials* 16, 951–956, 1995.
31. He, W. and Bellamkonda, R. V. Nanoscale neuro-integrative coatings for neural implants. *Biomaterials* 26, 2983–2990, 2005.
32. Huber, M. et al. Modification of glassy carbon surfaces with synthetic laminin-derived peptides for nerve cell attachment and neurite growth. *Journal of Biomedical Materials Research* 41, 278–288, 1998.
33. James, C. D. et al. Patterned protein layers on solid substrates by thin stamp microcontact printing. *Langmuir* 14, 741–744, 1998.
34. Martin, D. C. and Tresco, P. A. Biomaterials for the Central Nervous System. Final Project Report for Contract N01-NS-1-2338, NINDS, NIH, 2006.
35. Humphrey, C. D. and Schmidt, E. M. *Extracellular Single-Unit Recording Methods.* Humana Press, Totowa, NJ, 1990.
36. Garell, P. C. et al. Introductory overview of research instruments for recording the electrical activity of neurons in the human brain. *Review of Scientific Instruments* 69, 4027–4037, 1998.
37. Rutten, W. L. C. Selective electrical interfaces with the nervous system. *Annual Review of Biomedical Engineering* 4, 407–452, 2002.
38. Merrill, D. R., Bikson, M., and Jefferys, J. G. R. Electrical stimulation of excitable tissue: design of efficacious and safe protocols. *Journal of Neuroscience Methods* 141, 171–198, 2005.
39. Bierer, S. B. Optimization of Multi-Neuron Recordings Using Micro-Machined Electrode Arrays. Ph.D. thesis, Biomedical Engineering Department, University of Michigan, Ann Arbor, 2001.
40. Henze, D. A. et al. Intracellular features predicted by extracellular recordings in the hippocampus *in vivo. Journal of Neurophysiology* 84, 390–400, 2000.
41. Goepel, W. Ultimate limits in the miniaturization of chemical sensors. *Sensors and Actuators A: Physical* 56, 83–102, 1996.
42. Athreya, S. and Martin, D. C. Impedance spectroscopy of protein polymer modified silicon micromachined probes. *Sensors and Actuators A: Physical* 72, 203–216, 1999.
43. Buchko, C. J., Chen, L. C., Shen, Y., and Martin, D. C. Processing and microstructural characterization of porous biocompatible protein polymer thin films. *Polymer* 40, 7397–7407, 1999.
44. Kenaway, E.-R. et al. Release of tetracycline hydrochloride from electrospun poly(ethylene-co-vinylacetate), poly(lactic acid), and a blend. *Journal of Controlled Release* 81, 57–64, 2002.
45. Luu, Y. K., Kim, K., Hsiao, B. S., Chu, B., and Hadjiargyrou, M. Development of a nanostructured DNA delivery scaffold via electrospinning of PLGA and PLA-PEG block copolymers. *Journal of Controlled Release* 89, 341–353, 2003.

46. Sadik, O. A. and Wallace, G. G. Effect of polymer composition on the detection of electroinactive species using conductive polymers. *Electroanalysis* 5, 555–563, 1993.
47. Trojanowicz, M. and Krawczyk, T. K. V. Electrochemical biosensors based on enzymes immobilized in electropolymerized films. *Mikrochimica Acta* 121, 167–181, 1995.
48. Trojanowicz, M. and Krawczyk, T. K. V. Organic conducting polymers as active materials in electrochemical chemo-sensors and biosensors. *Chemia Analityczna* 42, 199–213, 1997.
49. Situmorang, M., Gooding, J. J., Hibbert, D. B., and Barnett, D. Electrodeposited polytyramine as an immobilisation matrix for enzyme biosensors. *Biosensors and Bioelectronics* 13, 953–962, 1998.
50. Gerard, M., Chaubey, A., and Malhotra, B. D. Application of conducting polymers to biosensors. *Biosensors and Bioelectronics* 17, 345–359, 2002.
51. Gao, M., Dai, L., and Wallace, G. G. Glucose sensors based on glucose-oxidase-containing polypyrrole/aligned carbon nanotube coaxial nanowire electrodes. *Synthetic Metals* 137, 1393–1394, 2003.
52. Nguyen, T. P., Le Rendu, P., Long, P. D., and De Vos, S. A. Chemical and thermal treatment of PEDOT:PSS thin films for use in organic light emitting diodes. *Surface and Coatings Technology* 180-181, 646–649, 2004.
53. Nagai, H. and Segawa, H. Energy-storable dye-sensitized solar cell with a polypyrrole electrode. *Chemical Communications* 8, 974–975, 2004.
54. Kurian, M., Galvin, M. E., Trapa, P. E., Sadoway, D. R., and Mayes, A. M. Single-ion conducting polymer-silicate nanocomposite electrolytes for lithium battery applications. *Electrochimica Acta* 50, 2125–2134, 2005.
55. Granqvist, C. G. Electrochromic tungsten oxide films: Review of progress 1993–1998. *Solar Energy Materials and Solar Cells* 60, 201–262, 2000.
56. Smith, A. B. and Knowles, C. J. Potential role of a conducting polymer in biochemistry—protein-binding properties. *Biotechnology and Applied Biochemistry* 12, 661–669, 1990.
57. Kanazawa, K. K. et al. "Organic metals": Polypyrrole, a stable synthetic "metallic" polymer. *Chemical Communications* 19, 854–855, 1979.
58. Schmidt, C. E., Shastri, V. R., Vacanti, J. P., and Langer, R. Stimulation of neurite outgrowth using an electrically conducting polymer. *Proceedings of the National Academy of Sciences USA* 94, 8948–8953, 1997.
59. George, P. M. et al. Fabrication and biocompatibility of polypyrrole implants suitable for neural prosthetics. *Biomaterials* 26, 3511–3519, 2005.
60. Nyberg, T., Inganäs, O., and Jerregard, H. Polymer hydrogel microelectrodes for neural communication. *Biomedical Microdevices* 4(1), 43–52, 2002.
61. Asberg, P. and Inganäs, O. Hydrogels of a conducting conjugated polymer as 3-D enzyme electrode. *Biosensors & Bioelectronics* 19, 199–207, 2003.
62. Llinas, R. R., Walton, K. D., Nakao, M., Hunter, I., and Anquetil, P. A. Neuro-vascular central nervous recording/stimulating system: Using nanotechnology probes. *Journal of Nanoparticle Research* 7, 111–127, 2005.
63. Cosnier, S. Biomolecule immobilization on electrode surfaces by entrapment or attachment to electrochemically polymerized films. A review. *Biosensors & Bioelectronics* 14, 443–456, 1999.
64. Trojanowicz, M., Matuszewski, W., and Podsiadia, M. Enzyme entrapped polypyrrole modified electrode for flow-injection determination of glucose. *Biosensors & Bioelectronics* 5, 149–156, 1990.

65. Piro, B., Dang, L. A., Pham, M. C., Fabiano, S., and Tran-Minh, C. A glucose biosensor based on modified-enyzme incorporated within electropolymerised poly(3,4-ethylenedioxythiophene) (PEDT) films. *Journal of Electroanalytical Chemistry* 512, 101–109, 2001.

66. Barisci, J. N., Hughes, D., Minett, A., and Wallace, G. G. Characterisation and analytical use of a polypyrrole electrode containing anti-human serum albumin. *Analytica Chimica Acta* 371, 39–48, 1998.

67. Bidan, G. M. et al. Electropolymerization as a versatile route for immobilizing biological species onto surfaces—Application to DNA biochips. *Applied Biochemistry and Biotechnology* 89, 183–193, 2000.

68. Wallace, G. G., Spinks, G. M., and Teasdale, P. R. *Conductive Electroactive Polymers: Intelligent Materials Systems.* CRC Press, Boca Raton, FL, 2003.

69. Yamato, H., Ohwa, M., and Wernet, W. Stability of polypyrrole and poly(3,4-ethylenedioxythiophene) for biosensor application. *Journal of Electroanalytical Chemistry* 397, 163–170, 1995.

70. Heywang, G. and Jonas, F. Poly(alkylenedioxythiophene)s—New, very stable conducting polymers. *Advanced Materials* 4, 116–118, 1992.

71. Cui, X. and Martin, D. C. Electrochemical deposition and characterization of poly(3,4-ethylenedioxythiophene) on neural microelectrode arrays. *Sensors and Actuators B: Chemical* 89, 92–102, 2003.

72. Xiao, Y. et al. Electrochemical polymerization of poly(hydroxymethylated-3,4-ethylenedioxythiophene) (PEDOT-MeOH) on multichannel neural probes. *Sensors and Actuators B: Chemical* 99, 437–443, 2004.

73. Xiao, Y., Cui, X., and Martin, D. C. Electrochemical polymerization and properties of PEDOT/S-PEDOT on neural microelectrode arrays. *Journal of Electroanalytical Chemistry* 573, 43–48, 2004.

74. Xiao, Y., Martin, D. C., Cui, X., and Shenai, M. Surface modification of neural probes with conducting polymer poly(hydroxymethylated-3,4-ethylenedioxythiophene) and its biocompatibility. *Applied Biochemistry and Biotechnology* 128, 117–129, 2006.

75. Yang, J., Xiao, Y., and Martin, D. C. Electrochemical polymerization of conducting polymer coatings on neural prosthetic devices: Nanomushrooms of polypyrrole using block copolymer films as templates. *Materials Research Society* 734, B8.4.1–11, 2003.

76. Thomas, C. A., Zong, K., Schottland, P., and Reynolds, J. R. Poly(3,4-alkylenedioxypyrrole)s as highly stable aqueous-compatible conducting polymers with biomedical implications. *Advanced Materials* 12, 222–225, 2000.

77. Yang, J., Kim, D., Hendricks, J., and Martin, D. C. Ordered surfactant-templated poly(3,4-ethylenedioxythiophene) (PEDOT) conducting polymer on microfabricated neural probes. *Acta Biomaterialia* 1, 124–136, 2005.

78. Hulvat, J. F. and Stupp, S. I. Liquid crystal templating of conducting polymers. *Angewandte Chemie-International Edition* 42, 778–781, 2003.

79. Cui, X., Hetke, J. F., Wiler, J. A., Anderson, D. J., and Martin, D. C. Electrochemical deposition and characterization of conducting polymer polypyrrole/PSS on multichannel neural probes. *Sensors and Actuators A: Physical* 93, 8–18, 2001.

80. Yang, J. and Martin, D. C. Impedance spectroscopy and nanoindentation of conducting PEDOT coatings on neural prosthetic devices. *Journal of Materials Research* 21, 1124–1132, 2006.

81. Maynard, E. M., Nordhausen, C. T., and Normann, R. A. The Utah Intracortical Electrode Array: A recording structure for potential brain-computer interfaces. *Electroencephalography and Clinical Neurophysiology* 102, 228–239, 1997.

82. Branner, A., Stein, R. B., and Normann, R. A. Selective stimulation of cat sciatic nerve using an array of varying-length microelectrodes. *Journal of Neurophysiology* 85, 1585–1594, 2001.

83. Ramanathan, K., Ram, M. K., Malhotra, B. D., and Murthy, S. N. Application of poly-aniline-Langmuir-Blodgett films as a glucose biosensor. *Materials Science and Engineering: C Biomimetic Materials Sensors and Systems* 3, 159–163, 1995.

84. Gambhir, A., Gerard, M., Mulchandani, A. K., and Malhotra, B. D. Coimmobilization of urease and glutamate dehydrogenase in electrochemically prepared polypyrrole-polyvinyl sulfonate films. *Applied Biochemistry and Biotechnology* 96, 249–257, 2001.

85. Chaubey, A., Gerard, M., Singhal, R., Singh, V. S., and Malhotra, B. D. Immobilization of lactate dehydrogenase on electrochemically prepared polypyrrole-polyvinyl-sulphonate composite films for application to lactate biosensors. *Electrochimica Acta* 46, 723–729, 2000.

86. Ramanathan, K. et al. Covalent immobilization of glucose oxidase to poly(o-amino benzoic acid) for application to glucose biosensor. *Journal of Applied Polymer Science* 78, 662–667, 2000.

87. Garner, B., Georgevich, A., Hodgson, A., Liu, L., and Wallace, G. G. Polypyrrole-heparin composites as stimulus-responsive substrate for endothelial cell growth. *Journal of Biomedical Materials Research* 44, 121–129, 1999.

88. Zhou, D., Too, C. O., and Wallace, G. G. Synthesis and characterisation of polypyrrole/heparin composites. *Reactive and Functional Polymers* 39, 19–26, 1999.

89. Collier, J. H., Camp, J. P., Hudson, T. W., and Schmidt, C. E. Synthesis and characterization of polypyrrole-hyaluronic acid composite biomaterials for tissue engineering applications. *J. Biomed. Mater. Res.* 50, 574–584, 2000.

90. Kim, D.-H., Richardson-Burns, S. M., Hendricks, J., Sequera, C., and Martin, D. C. Effect of immobilized nerve growth factor (NGF) on conductive polymers: Electrical properties and cellular response. *Advanced Functional Materials* 17, 79–86, 2006.

91. Draget, K. I., Ostgaard, K., and Smidsrod, O. Homogeneous alginate gels—A technical approach. *Carbohydrate Polymers* 14, 159–178, 1990.

92. Smetana, K. Cell biology of hydrogels. *Biomaterials* 14, 1046–1050, 1993.

93. Smidsrod, O. and Skjakbraek, G. Alginate as immobilization matrix for cells. *Trends in Biotechnology* 8, 71–78, 1990.

94. Kuo, C. K. and Ma, P. X. Ionically crosslinked alginate hydrogels as scaffolds for tissue engineering. Part 1. Structure, gelation rate and mechanical properties. *Biomaterials* 22, 511–521, 2001.

95. LeRoux, M. A., Guilak, F., and Setton, L. A. Compressive and shear properties of alginate gel: Effects of sodium ions and alginate concentration. *Journal of Biomedical Materials Research* 47, 46–53, 1999.

96. Lee, K. Y., Peters, M. C., Anderson, K. W., and Mooney, D. J. Controlled growth factor release from synthetic extracellular matrices. *Nature* 408, 998–1000, 2000.

97. Tabata, Y., Matsui, Y., and Ikada, Y. Growth factor release from amylopectin hydrogel based on copper coordination. *Journal of Controlled Release* 56, 135–148, 1998.

98. Celebi, N., Erden, N., Gonul, B., and Koz, M. Effects of epidermal growth-factor dosage forms on dermal wound strength in mice. *Journal of Pharmacy and Pharmacology* 46, 386–387, 1994.

99. Winn, S. R., Uludag, H., and Hollinger, J. O. Sustained release emphasizing recombinant human bone morphogenetic protein-2. *Advanced Drug Delivery Reviews* 31, 303–318, 1998.

100. Kim, D.-H. and Martin, D. C. Sustained release of dexamethasone from hydrophilic matrices using PLGA nanoparticles for neural drug delivery. *Biomaterials* 27, 3031–3037, 2006.

101. Biran, R., Martin, D. C., and Tresco, P. A. The brain tissue response to implanted silicon microelectrode arrays is increased when the device is tethered to the skull. *Journal of Biomedical Materials Research A* (published online) doi 10.1002/jbm.a.31135, 2007.

102. Gheith, M. K., Sinani, V. A., Wicksted, J. P., Matts, R. L., and Kotov, N. A. Single-walled carbon nanotube pollyelectrolyte multilayers and freestanding films as a biocompatible platform for neuroprosthetic implants. *Advanced Materials* 17, 2663–2670, 2005.

103. Nguyen-Vu, T. D. B. et al. Vertically aligned carbon nanofiber arrays: An advance toward electrical-neural interfaces. *Small* 2, 89–94, 2006.

104. Arnold, M. S., Stupp, S. I., and Hersam, M. C. Enrichment of single-walled carbon nanotubes by diameter in density gradients. *Nano Letters* 5, 713–718, 2005.

105. Kmecko, T. et al. Nanocomposites for Neural Interfaces. Paper #0926-CC04-06, Materials Research Society, 2006.

106. Smela, E. Conjugated polymer actuators for biomedical applications. *Advanced Materials* 15, 481–494, 2003.

107. Gandhi, M. R., Murray, P., Spinks, G. M., and Wallace, G. G. Mechanism of electromechanical actuation in polypyrrole. *Synthetic Metals* 73, 247–256, 1995.

108. Baughman, R. H. Conducting polymer artificial muscles. *Synthetic Metals* 78, 339–353, 1996.

109. Pei, Q. B. and Inganas, O. Conjugated polymers and the bending cantilever method: Electrical muscles and smart devices. *Advanced Materials* 4, 277–278, 1992.

110. Otero, T. F. and Cortes, M. T. Artificial muscles with tactile sensitivity. *Advanced Materials* 15, 279–282, 2003.

111. Polikov, V. S., Tresco, P. A., and Reichert, W. A. Response of brain tissue to chronically implanted neural electrodes. *Journal of Neuroscience Methods* 148, 1–18, 2005.

112. Turner, J. N. et al. Cerebral astrocyte response to micromachined silicon implants. *Experimental Neurology* 156, 33–49, 1999.

113. Vetter, R. J., Williams, J. C., Hetke, J. F., Nunamaker, E. A., and Kipke, D. R. Chronic neural recording using silicon-substrate microelectrode arrays implanted in cerebral cortex. *IEEE Transactions on Biomedical Engineering* 51, 896–904, 2004.

114. Williams, J. C., Rennaker, R. L., and Kipke, D. R. Long-term neural recording characteristics of wire microelectrode arrays implanted in cerebral cortex. *Brain Research: Brain Research Protocols* 4, 303–313, 1999.

115. Jenkner, M., Muller, B., and Fromherz, P. Interfacing a silicon chip to pairs of snail neurons connected by electrical synapses. *Biological Cybernetics* 84, 239–249, 2001.

116. Fromherz, P. Electrical interfacing of nerve cells and semiconductor chips. *Chemphyschem* 3, 276–284, 2002.

117. Cui, D. and Gao, H. Advance and prospect of bionanomaterials. *Biotechnology Progress* 19, 683–692, 2003.

118. Drotleff, S. et al. Biomimetic polymers in pharmaceutical and biomedical sciences. *European Journal of Pharmaceutics and Biopharmaceutics* 58, 385–407, 2004.

119. Williams, D. Biomimetic surfaces: How man-made becomes man-like. *Medical Device Technology* 6, 6–10, 1995.

120. Brahim, S., Wilson, A. M., Narinesingh, D., Iwuoha, E., and Guiseppi-Elie, A. Chemical and biological sensors based on electrochemical detection using conducting electroactive polymers. *Microchimca Acta* 143, 123–137, 2003.

121. Schmidt, C. E. and Leach, J. B. Neural tissue engineering: Strategies for repair and regeneration. *Annual Review of Biomedical Engineering* 5, 293–347, 2003.

122. Buchko, C. J., Kozloff, K. M., and Martin, D. C. Surface characterization of porous, biocompatible protein polymer thin films. *Biomaterials* 22, 1289–1300, 2001.

123. Zhong, Y. H. and Bellamkonda, R. V. Controlled release of anti-inflammatory agent alpha-MSH from neural implants. *Journal of Controlled Release* 106, 309–318, 2005.

124. Campbell, T. E., Hodgson, A. J., and Wallace, G. G. Incorporation of erythrocytes into polypyrrole to form the basis of biosensor to screen for rhesus (D) blood groups and rhesus (D) antibodies. *Electroanalysis* 11, 215–222, 1999.

125. Shi, Y. Mechanisms of caspase activation and inhibition during apoptosis. *Mol. Cell* 9, 459–470, 2002.

126. Friedlander, R. M. Apoptosis and capsases in neurodegenerative diseases. *New England Journal of Medicine* 348, 635–-643, 2003.

127. Zimerman, B., Volberg, T., and Geiger, B. Early molecular events in the assembly of the focal adhesion-stress fiber complex during fibroblast spreading. *Cell Motility and the Cytoskeleton* 58, 143–159, 2004.

128. Bershadsky, A. D., Balaban, N. Q., and Geiger, B. Adhesion-dependent cell mechanosensitivity. *Annual Review of Cell and Developmental Biology* 19, 677–695, 2003.

129. Wozniak, M. A., Modzelewska, K., Kwong, L., and Keely, P. J. Focal adhesion regulation of cell behavior. *Biochimica et Biophysica Acta—Molecular Cell Research* 1692, 103–119, 2004.

130. Bijian, K. et al. Actin cytoskeleton regulates extracellular matrix-dependent survival signals in golmerular epithelial cells. *American Journal of Physiology—Renal Physiology* 289, F1313–F1323, 2005.

131. Liu, J. and Lin, A. N. Role of JNK activation in apoptosis: A double-edged sword. *Cell Research* 15, 36–42, 2005.

132. Reddig, P. J. and Juliano, R. L. Clinging to life: cell to matrix adhesion and cell survival. *Cancer and Metastasis Reviews* 24, 425–439, 2005.

133. Allani, P. K., Sum, T., Bhansali, S. G., Mukherjee, S. K., and Sonee, M. A comparative study of the effect of oxidative stress on the cytoskeleton in human cortical neurons. *Toxicology and Applied Pharmacology* 196, 29–36, 2004.

134. Groenendall, L. B., Jonas, F., Freitag, D., Pielartzik, H., and Reynolds, J. R. Poly(3,4-ethylenedioxythiophene) and its derivatives: Past, present, and future. *Advanced Materials* 12, 481–494, 2000.

135. Xing, K. Z., Fahlman, M., Chen, X. W., Inganäs, O., and Salaneck, W. R. The electronic structure of poly(3,4-ethylenedioxythiophene): Studied by XPS and UPS. *Synthetic Metals* 89, 161–165, 1997.

136. Beamson, G. and Briggs, D. *High Resolution XPS of Organic Polymers: The Scienta ESCA300 Database*. John Wiley & Sons, New York, 1992.

137. Kim, T. Y., Park, C. M., Kim, J. E., and Suh, K. S. Electronic, chemical and structural change induced by organic solvents in tosylate-doped poly(3,4-ethylenedioxythiophene) (PEDOT-OTs). *Synthetic Metals* 149, 169–174, 2005.

138. Greczynski, G., Kugler, T., and Salaneck, W. R. Characterization of the PEDOT-PSS system by means of X-ray and ultraviolet photoelectron spectroscopy. *Thin Solid Films* 354, 129–135, 1999.

139. Greczynski, G. et al. Photoelectron spectroscopy of thin films of PEDOT-PSS conjugated polymer blend: A mini-review and some new results. *Journal of Electron Spectroscopy and Related Phenomena* 121, 1–17, 2001.

140. Zotti, G. et al. Electrochemical and XPS studies toward the role of monomers and polymeric sulfonate counterions in the synthesis, composition, and properties of poly(3,4-ethylenedioxythiophene). *Macromolecules* 36, 3337–3344, 2003.

141. Crispin, X. et al. Conductivity, morphology, interfacial chemistry, and stability of poly(3,4-ethylenedioxythiophene)-poly(styrene sulfonate): A photoelectron spectroscopy study. *Journal of Polymer Science: Part B: Polymer Physics* 41, 2561–2583, 2003.

142. Marciniak, S. et al. Light induced damage in poly(3,4-ethylenedioxythiophene) and its derivatives studied by photoelectron spectroscopy. *Synthetic Metals* 141, 67–73, 2004.

143. Denier van der Gon, A. W., Birgerson, J., Fahlman, M., and Salaneck, W. R. Modification of PEDOT-PSS by low energy electrons. *Organic Electronics* 3, 111–118, 2002.

144. Wang, T., Qi, Y., Xu, J., Hu, X., and Chen, P. Effects of poly(ethylene glycol) on electrical conductivity of poly(3,4-ethylenedioxythiophene)-poly(styrenesulfonic acid) film. *Applied Surface Science* 250, 188–194, 2005.

145. Paczosa-Bator, B., Peltonen, J., Bobacka, J., and Lewenstam, A. Influence of morphology and topography on potentiometric response of magnesium and calcium sensitive PEDOT films doped with adenosine triphosphate (ATP). *Analytica Chimica Acta* 555, 118–127, 2006.

146. Li, X., Li, Y., Tan, Y., Yang, C., and Li, Y. Self-assembly of gold nanoparticles prepared with 3,4-ethylenedioxythiophene as reductant. *Journal of Physical Chemistry B* 108, 5192–5199, 2004.

147. Khan, M. A. et al. Surface characterization of poly(3,4-ethylenedioxythiophene)-coated latexes by X-ray photoelectron spectroscopy. *Langmuir* 16, 4171–4179, 2000.

148. Jönsson, S. K. M., Salaneck, W. R., and Fahlman, M. Spectroscopy of ethylenedioxythiophene-derived systems: From gas phase to surfaces and interfaces found in organic electronics. *Journal of Electron Spectroscopy and Related Phenomena* 137–140, 805–809, 2004.

149. Nguyen, T. P. and de Vos, S. A. An investigation into the effect of chemical and thermal treatments on the structural changes of poly(3,4-ethylenedioxythiophene)/polystyrenesulfonate and consequences on its use on indium tin oxide. *Applied Surface Science* 221, 330–339, 2004.

150. Fedorow, H. et al. Neuromelanin in human dopamine neurons: Comparison with peripheral melanins and relevance to Parkinson's disease. *Progress in Neurobiology* 75, 109–124, 2005.

151. Gonzalez, R., Sanchez, A., Chicarro, M., Rubianes, M. D., and Rivas, G. A. Dopamine and glucose sensors based on glassy carbon electrodes modified with melanic polymers. *Electroanalysis* 16, 1244–1253, 2004.

152. Rubianes, M. D. and Rivas, G. A. Highly selective dopamine quantification using a glassy carbon electrode modified with a melanin-type polymer. *Analytica Chimica Acta* 440, 99–108, 2001.

153. Subianto, S., Will, G., and Meredith, P. Electrochemical synthesis of melanin freestanding films. *Polymer* 46, 11505–11509, 2005.

154. Zielinski, M. and Pande, C. Electrochemical synthesis and properties of poly(dihydroxyindole), a synthetic analog of melanin. *Synthetic Metals* 37, 350–351, 1990.

155. Horak, V. and Weeks, G. Poly(5,6-dihydroxyindole) melanin film electrode. *Bioorganic Chemistry* 21, 24–33, 1993.

156. Gidanian, S. and Farmer, P. J. Redox behavior of melanins: Direct electrochemistry of dihydroxyindole-melanin and its Cu and Zn adducts. *Journal of Inorganic Biochemistry* 89, 54–60, 2002.
157. Stark, K. B. et al. Effect of stacking and redox state on optical absorption spectra of melanins—Comparison of theoretical and experimental results. *Journal of Physical Chemistry B* 109, 1970–1977, 2005.
158. Capozzi, V. et al. Optical and photoelectronic properties of melanin. *Thin Solid Films* 511, 362–366, 2006.
159. Jastrzebska, M. M., Isotalo, H., Paloheimo, J., Stubb, H., and Pilawa, B. Effect of Cu2+-ions on semiconductor properties of synthetic DOPA melanin polymer. *Journal of Biomaterials Science—Polymer Edition* 7, 781–793, 1996.
160. Osak, W., Tkacz, K., Czternastek, H., and Slawinski, J. I.-V. Characteristics and electrical conductivity of synthetic melanin. *Biopolymers* 28, 1885–1890, 1989.
161. Meredith, P. et al. Toward structure-property-function relationships for eumelanin. *Soft Matter* 2, 37–44, 2006.
162. Tran, L. M., Powell, B. J., and Meredith, P. Chemical and structural disorder in eumelanins: A possible explanation for broadband absorbance. *Biophysical Journal* 90, 743–752, 2006.
163. Ostergren, A., Svensson, A. L., Lindquist, N. G., and Brittebo, E. B. Dopamine melanin-loaded PC12 cells: A model for studies on pigmented neurons. *Pigment Cell Research* 18, 306–314, 2005.
164. Li, J. et al. Differential effects of human neuromelanin and synthetic dopamine melanin on neuronal and glial cells. *Journal of Neurochemistry* 95, 599–608, 2005.
165. Richardson-Burns, S. M. et al. Polymerization of the conducting polymer poly(3,4-ethylenedioxythiophene) (PEDOT) around living neural cells. *Biomaterials* 28(8) 1539–1552, 2007.
166. Richardson-Burns, S. M. et al. Electrochemical polymerization of conducting polymers in living neural tissue. *Journal of Neural Engineering* 4(2), 26–213, 2007.
167. Jonnson, S. K. M. et al. The effects of solvents on the morphology and sheet resistance in PEDOT-PSS films. *Synthetic Metals* (13P), 1–10, 2003.
168. Richardson, R. T., Thompson, B., Moulton, S., Newbold, C., Lum, M. G., Cameron, A. et al. The effect of polypyrrole with incorporated neurotrophin-3 on the promotion of neurite outgrowth from auditory neurons. *Biomaterials* 28, 513–523, 2007.

8 Strategies for Regeneration and Repair in the Injured Central Nervous System

Molly S. Shoichet, Ciara C. Tate,
M. Douglas Baumann, and Michelle C. LaPlaca

CONTENTS

8.1 INTRODUCTION

In this chapter we review the consequences of traumatic injury to the brain and spinal cord and the regenerative strategies that are being pursued to overcome the tissue degeneration that ensues. The chapter begins with a discussion of the epidemiology and pathogenesis of traumatic injury and is followed by a discussion of the challenges associated with promoting regeneration and limiting degeneration. Strategies for treatment are described including drug delivery, cell-based, and tissue engineering approaches aimed at achieving functional recovery after traumatic central nervous system (CNS) injury. The chapter concludes with an outlook on research dedicated to future treatment strategies.

8.1.1 TRAUMATIC INJURY IN THE CENTRAL NERVOUS SYSTEM (CNS): BRAIN AND SPINAL CORD

The brain and spinal cord make up the CNS, with the brain coordinating higher-level functions and the spinal cord serving mainly as the communication pathway between the brain and the periphery. Traumatic brain injury (TBI) and traumatic spinal cord injury (SCI) result when an external physical insult causes damage, which can range from mild to severe. Disabilities from CNS trauma are a function of the mode, severity, and anatomical location of the insult. Although a mechanical insult is the initiating factor in the pathophysiology of CNS trauma, the distribution of the force(s) and the resulting cellular responses are currently not well understood. Both the brain and spinal cord are composed of gray matter (consisting of neuronal cell bodies) and white matter (myelinated axons). The outer layer of the brain consists of gray matter, and therefore blunt traumatic impact to the brain often consists of a large degree of cell body damage, while high-speed impacts also include axonal damage in deeper regions. In contrast, the outer layer of the spinal cord consists of axonal tracts, and primary damage from compression and blunt impact commonly includes damage to both gray and white matter, especially given the relatively smaller thickness of the spinal cord compared to the brain. Regardless of the initial location of the insult to the CNS, injury is an ongoing process, with primary damage leading to a cascade of deleterious events that can affect both cell body and axonal function, resulting in continued dysfunction and prolonged degeneration.

TBI often leads to motor (e.g., loss of ambulation, balance, coordination, fine motor skills, strength, and endurance) and cognitive (e.g., loss of communication, information processing, memory, and perceptual skills) impairment [1–11]. Focal injuries, which are typically induced when an object impacts the skull (such as during a blow to the head, motor vehicle accident, or fall), result from direct loading and can occur without diffuse damage. As a result, macroscopic damage is typically visible at the site of impact, and the clinical symptoms are specific to the anatomical area that is injured. Contact loading can also result in coup (at the site of impact) and contra-coup (away from the site of impact) contusions to the brain, involving both cellular and vascular components. Focal injuries account for half of all severe head injuries, but two-thirds of all deaths are in this group [12,13]. Clinically, TBI severity is rated using the Glasgow coma score, in which assessments of eye, verbal, and motor responses give a combined score ranging from 3 (deep coma or dead) to 15 (fully alert). A Glasgow coma score ≤8 is considered severe, 9 to 12 is moderate, and ≥13 is minor [14]. Clinical responses aim at stabilizing the patient, though this has only a modest effect on decreasing the propagation of cell and tissue damage and does not promote repair or regeneration. Most experimental therapies have been developed to decrease secondary damage with pharmacological agents, but these attempts are limited in their ability to improve clinical outcomes [15–20] . Ideally, clinical therapies for TBI will significantly reduce secondary cell damage and death by targeting a variety of mechanisms and enhance regeneration and plasticity for improved functional recovery.

Traumatic SCI results from either compression, where the vertebral column impinges on it, or from transection (such as from a stab or gunshot wound). Severity of injury is ranked on the American Spinal Injury Association scale of A to E, where A is a complete SCI with no motor or sensory function in the perineal region, and B through E reflect improving motor function below the site of injury. This definition is relatively new and is one reason why the reported ratio of complete to incomplete injuries varies. In the United States approximately 60% of spinal cord injuries are complete, while studies from Canada and Brazil report complete injuries in 18 and 87% of cases, respectively [21]. Neuroregeneration after SCI requires axonal outgrowth of existing and new axons plus myelination, ultimately leading to functional recovery. To this end, researchers have focused on delivering either factors, such as neurotrophins, that act on regeneration-associated genes to stimulate axonal sprouting and growth or molecules to neutralize the inhibitory cues present in the injured spinal cord.

Degeneration in the CNS as a result of disease or trauma can have devastating effects on quality of life. Each year sees 1.4 million new cases of TBI in the United States, resulting in over 50,000 deaths and 80,000 disabilities [22,23] and 10,000 new cases of SCI, with another 275,000 people living with an SCI [24]. Most of the injuries are sustained by young men and women, a grim statistic compounded by the lifelong disability that usually results. These statistics, combined with negative health and sociological implications and an economic impact of over $70 billion directly related to CNS injury [24,25] prompt the demand for novel and clinically effective treatments. The complexity of the CNS and variability in clinical presentation make efforts to repair damaged CNS tissue and restore function particularly challenging. However, recent progress in regenerative medicine has led to new approaches for the treatment of CNS injuries and thus renewed hope.

8.2   ENVIRONMENT OF THE INJURED CNS: CHALLENGES AND OPPORTUNITIES

The evolution of the neural injury response involves a dynamic interplay between events promoting repair and regeneration and those of damage and inhibition (see Figure 8.1 for an overview). Physical insults to brain or spinal cord tissue initially cause necrotic cell death and damage in the underlying tissue, followed by multiple subsequent events such as opening of the blood–brain or blood–spinal cord barrier, inflammation, edema, ischemia, excitotoxicity, increase in free radicals, and altered cell signaling and gene expression (for reviews, see [26,27]). This cascade of secondary events leads to additional cell death, demyelination, axonal degeneration, formation of a glial scar, and an inhibitory environment for regeneration [28].

8.2.1   BLOOD–BRAIN BARRIER (BBB) AND BLOOD–SPINAL CORD BARRIER (BSCB)

The BBB and the BSCB are similar and play an important role in the normal and pathologic injury response. These barriers consist of closely packed endothelial

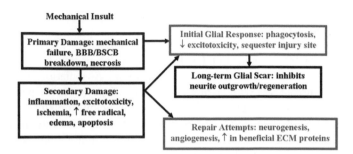

FIGURE 8.1 Schematic of events following focal traumatic CNS injury. Mechanisms of damage and inhibition of regeneration (in black boxes) occur alongside permissive and reparative mechanisms (in gray boxes). Better understanding mechanisms on both sides of the response can aid in the development of more effective clinical treatments for traumatic injury to the CNS.

cells surrounded by astrocytic processes that protect the brain and spinal cord from biochemical fluctuations in the periphery. A breach in the BBB or BSCB permits infiltration of macrophages, fibroblasts and other cell types, and inflammatory molecules into the site of injury. This activates astrocytes, contributing to the inhibitory environment of the reactive glial scar, which physically and chemically obstruct axonal regeneration (see Section 8.2.3). The opening of and damage to the BBB and BSCB is a hallmark of CNS injury and disease, and while it can expose neural tissue to harmful blood-borne components, this symptom can also be exploited for delivery of neuroprotective and reparative molecules.

8.2.2 Dual Role of Inflammation in Traumatic CNS Injury

The inflammatory response is a subject of active debate within the neuroscience community. While some inflammation is clearly needed to limit degeneration and address the cellular debris resulting from CNS injury, there is active discussion on whether the inflammatory response should be further enhanced [29–32]. Proinflammatory cytokines are released within minutes following traumatic injury, with acutely harmful effects (e.g., BBB and BSCB dysfunction, promotion of neuronal death) but are beneficial at later time points (e.g., inducing synthesis of antiinflammatory cytokines, inducing neurotrophic factor secretion, and promoting proliferation of oligodendrocyte precursor cells that may help in remyelination) [31]. Neurons and glial cells produce chemokines and complement proteins and have corresponding receptors. In a similarly dual role, chemokines and complement proteins are involved in acute BBB and BSCB dysfunction and edema development, but these proteins eventually lead to increased growth factor production. Furthermore, complement proteins have been found to protect neurons from excitotoxicity-induced apoptosis and promote opsonization [33].

With respect to the cellular aspect of the inflammatory response, microglia are the first to respond (minutes to hours) by proliferating, activating, and migrating to the area of injury, where they essentially function as macrophages [34,35]. Increased BBB and BSCB permeability contributes to leukocyte infiltration from the blood to the injury site, a process that is mediated by cytokines, chemokines, and complement proteins [33]. Neutrophils infiltrate (hours to days), followed by monocytes (days) [34]. Again, these immune cells have dual roles. The oxidative burst of neutrophils and macrophages is harmful because of the release of oxygen free radicals and neurotoxic enzymes [33]; however, both activated microglia and monocyte-derived macrophages aid in clearing debris from dead and damaged cells via phagocytosis [36].

Because of the dual nature of the inflammatory response, treatments for TBI and SCI that target specific cells or proteins involved in the inflammatory response must be approached with caution. Schwartz has found that the injection of macrophages to the injured spinal cord can be beneficial [37], yet clinical trials have been stopped. Others question the utility of an increased inflammatory response, as Schwann cells seem to provide most of the clean-up after SCI rather than macrophages and microglia [38–40]. However, while Schwann cells have shown some benefit, on their own they are insufficient to promote functional improvement in SCI [41], in part because Schwann cells do not myelinate CNS axons similarly to oligodendrocytes [42].

8.2.3 GLIAL SCAR LIMITS REGENERATION

As part of the acute injury response, activated glial cells (including astrocytes and microglia) migrate to the injury site, where they form a tight and interpenetrating network known as the reactive glial scar. Initially, the beneficial role of the glial response helps buffer excitotoxic and cytotoxic molecules, repair the BBB and BSCB, and isolate the site of injury [43]. However, glial cells that persist at the injury site produce inhibitory factors that manifest within hours of the original insult and severely limit axonal regeneration, including myelin-associated glycoprotein [44,45], chondroitin sulfate proteoglycans [46], and Nogo [47], among others (for a review, see [48]). As part of a nascent regenerative response, severed CNS axons develop new growth cones that extend very short distances before stabilizing and then receding ("abortive sprouting") because of the inhibitory chemical environment [49].

To achieve axonal regeneration, which is particularly crucial following SCI, the physical and chemical inhibitory environment of the glial scar must be overcome. For example, Fawcett and colleagues found that the glial scar contains significant amounts of chondroitin sulphate proteoglycans and have investigated the use of an enzyme—chondroitinase ABC—to degrade the glial scar and thereby provide a pathway for regeneration with some success [50]. Others have found that the inhibitory environment presented by myelin following injury can be neutralized by targeting the Rho-kinase receptor [51,52]. Clinical trials for SCI are ongoing for the delivery of Cethrin®, a Rho antagonist, and for the delivery of anti-Nogo-A, a neutralizing molecule [53]. Interestingly, while similar inhibitory components are present after TBI, there is less effort to alter the environment imposed by the glial scar.

The complex environment of the traumatically injured brain and spinal cord exhibit aspects of inhibition and ongoing cell death, together with endogenous attempts at repair and regeneration. While significant efforts have been made to reduce inhibitory extracellular matrix expression following CNS injury, much less attention has been given to the role of endogenous reparative matrix proteins, such as fibronectin and laminin. Both of these extracellular matrix proteins are elevated following TBI and colocalize with macrophages and activated microglia (fibronectin) and brain microvasculature (laminin), suggesting that these proteins may play a role in repairing damaged brain tissue [54].

8.2.4 REGENERATION IS POSSIBLE FOLLOWING CNS INJURY

Until the 1980s, it was unclear whether CNS axons could regenerate; however, seminal studies by Aguayo and colleagues demonstrated that CNS axons could regenerate when provided with the appropriate environment. In these studies peripheral nerve grafts were implanted in the optic nerve, and axons were regenerated within the peripheral nerve graft but not beyond the graft and not into the tissue [55]. After traumatic injury to the brain or spinal cord, some neurogenesis is often observed within the first few months following injury. It has been appreciated recently that the injured brain may be attempting repair through developmental processes, as evidenced by increases in neurogenesis and angiogenesis that occur following TBI [56–58]. In SCI, neurogenesis is observed following surgical intervention where the pressure or compression exerted on the spinal cord by vertebrae is relieved by realigning these tissues [59].

These significant findings, coupled with others that have defined the inhibitory environment within the damaged CNS tissue, have guided a generation of regenerative strategies. While these recent breakthroughs offer promise in restoring function, a number of challenges remain because of the complex and dynamic milieu of the injured CNS. The evolution of the neural injury response involves a dynamic interplay between events promoting repair and regeneration and those of damage and inhibition. Thus, in addition to regenerative strategies, neuroprotective strategies seek to limit secondary injury (or the cascade of degenerative events) as a way to limit the loss of function that follows injury.

8.3 NEUROPROTECTION AND NEUROREGENERATION TREATMENT STRATEGIES

Efforts to treat traumatic CNS injures can be broadly divided into two categories: (a) neuroprotection, the minimization of cell damage and death and axonal degeneration caused by the cascade of secondary events, and (b) neuroregeneration, the promotion of plasticity and axonal growth. Given the anatomic and physiologic differences between the brain and spinal cord, strategies for neuroregeneration differ slightly (see Table 8.1). For neuroregeneration following traumatic SCI, axonal growth is required across the injury site, through the glial scar, and to the appropriate target tissue. Elucidating secondary damage events and exploiting factors involved in endogenous neuroprotection and neuroregeneration may aid in developing more effective treatments for traumatic CNS injury. Because of the progressive nature of cell death following traumatic CNS injury, a sustained neuroprotective therapy may be required to alleviate or reduce neurological disability and render the damaged CNS more receptive to regenerative strategies. Ideal treatment strategies will exploit and complement endogenous repair mechanisms while suppressing inhibitory mechanisms.

TABLE 8.1
Therapeutic Goals in TBI and SCI[a]

		Brain	Spinal Cord
Neuroprotection	Goal	↓ Cell dysfunction/death caused by secondary cascades	
	Examples	Antiapoptotic molecules, free radical scavengers, antiexcitotoxic molecules, trophic factors	
Neuroregeneration	Goal	↑ Plasticity (reorganization of neuronal circuits)	↑ Axonal growth to reconnect brain with target cells
	Examples	Trophic support, enhance neurogenesis	Trophic support, ↓ inhibitory components of glial scar, physical guidance channels with chemical cues

[a] Although their constituent cells are similar, therapeutic goals in the brain and spinal cord differ after traumatic injury. This is especially true in the case of complete SCI when few or no axons are spared, leaving axonal regeneration as the only regenerative goal.

8.3.1 DRUG DELIVERY TO THE INJURED CNS FOR NEUROPROTECTION AND NEUROREGENERATION

Drug delivery strategies have been investigated to both limit degeneration and promote regeneration following traumatic injury to the CNS. In TBI, current treatment methods in clinical practice primarily aim to reduce intracranial pressure in an effort to minimize brain damage caused by swelling. Examples include moderate hypocapnia and mannitol (first-line measures), followed by barbiturates, moderate hypothermia, or a decompressive craniectomy (second-line measures) if early attempts fail [60]. However, these therapies have a modest effect on acute brain damage and subsequent cell death pathways that lead to functional impairment [61]. Furthermore, these treatments do not provide sustained efforts that promote repair or regeneration.

Pharmacological approaches under investigation for TBI in the past several years target one or more of the pathological events following TBI. Excitotoxicity has been targeted in numerous animal models with promising results, yet various treatments that mediate associated cellular pathways, such as glutamate receptor antagonists, were not found to be effective for humans with TBI [62,63]. Similar results occurred with other investigational drugs including free radical scavengers and steroids [64]. These treatments may have failed in the clinic because they target pathways that are both deleterious and beneficial, so the dosage and time of treatment are critical, and perhaps require optimization, to avoid interfering with normal homeostasis and reparative mechanisms in the brain. Furthermore, these treatments targeted single mechanisms, which may be insufficient in light of the multifaceted pathology. Therapies that address multiple pathological events, such as hypothermia [65] or progesterone administration [66,67], are more promising and are currently being evaluated in clinical trials [68].

To limit degeneration after SCI, the goal has been to preserve functional behavior through axonal sparing and reduced lesion volumes. To date, only methylprednisolone, an antiinflammatory steroid, is used clinically following SCI, despite limited efficacy [69–72]. Other molecules that have been investigated by systemic delivery include GM-1 gangliosides [73], minocycline [74], and cAMP moderators [75]. GM-1 ganglioside, while failing its clinical trial, has been shown to inhibit apoptosis and promote axonal sprouting [76]. Numerous other drug molecules are under investigation, including those mentioned in Section 8.2.3.

The time course for intervention with neuroprotective strategies has typically been within 24 to 72 hours of traumatic injury, but pharmacological intervention is transient. Advances in sustained delivery of neuroprotective drugs have been made, but these approaches have been constrained by technical limitations. For example, the success of controlled pharmacological strategies using pumps in the brain has been limited by drug instability and poor release control [77–79]. Although controlled-release polymer systems can be designed to degrade in the body and release an encapsulated chemical with near zero-order kinetics [80–82], only a finite dose can be implanted, and multiple surgeries would be required for long-term release. Moreover, the diffusion distance is limited in both the brain and spinal cord, particularly for molecules for which there are extraneous receptors in adjacent tissue [83].

The route of drug delivery is also an important consideration. Molecules have been delivered systemically and injected locally, the latter via epidural, intrathecal, and intramedullary routes. Intramedullary injection can, however, cause tissue damage and has not been widely pursued. Systemic delivery is the easiest route and is suitable for molecules that can cross the BBB and BSCB; however, it often requires high doses with undesirable side effects. While the majority of pharmacological efforts for TBI have used systemic delivery, potential treatments for SCI have explored local delivery for molecules that cannot easily cross the BSCB.

Epidural drug delivery has been achieved with a continuous infusion of a drug solution [84], by injectable hydrogels [85], and through liposomes [86] for analgesic delivery. Epidural delivery is more localized than systemic delivery and relatively minimally invasive yet requires the therapeutic molecule to cross the dura mater, arachnoid mater, fluid-filled intrathecal space, and pia mater before penetrating the spinal cord, likely resulting in the loss of most of the dosage [87]. Notwithstanding these limitations, epidural delivery of Cethrin® delivered in a fibrin glue is currently in clinical trials.

Intrathecal delivery, most commonly pursued with a mini-pump and catheter system, requires the dura mater to be punctured but inherently achieves localized release at the injured tissue, requiring the therapeutic molecules to diffuse only across the pia mater and into the damaged tissue [88]. The mini-pump is implanted subcutaneously (as in Figure 8.2) and delivers a constant drug volume to the intrathecal space. This system is being pursued in an ongoing clinical trial for acute traumatic SCI using anti-Nogo-A [89]. Long-term use of catheters is, however, undesirable because of the possibility of infection, chronic inflammation, scarring, and compression of the spinal cord [90]. Injectable gels have also been used to localize drug delivery to the injured tissue and have been investigated in the epidural and

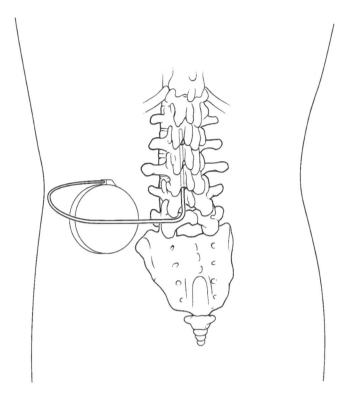

FIGURE 8.2   Implanted mini-pump with intraspinal catheter. Infusion mini-pumps may be implanted subcutaneously and an intraspinal catheter guided into the epidural or intrathecal space. These pumps provide good control of drug dosages but require surgery to implant and remove the pump. Long-term catheterization may cause scarring and compression of the spinal cord. (Copyright 2006. With permission from Michael Corrin.)

intrathecal spaces. Intrathecal delivery of injectable gels is being pioneered, as illustrated in Figure 8.3, with hydrogels and has been shown to be safe and effective for localized release [91–93].

8.3.2   CELL DELIVERY TO THE INJURED CNS FOR NEUROPROTECTION AND NEUROREGENERATION

Exploiting cells for transplantation offers great potential, with cells functioning as biologically active systems to produce specific beneficial factors or to replace lost cells and tissue. Though this section will focus on the use of donor cells, it should be noted that treatment strategies are also being developed to stimulate and augment endogenous stem cell populations. This strategy has shown promise following TBI [94], but this has not demonstrated functional benefit in SCI to date. Advantages of transplanting cells include the ability to target multiple neuroprotective and neuroregenerative mechanisms and the ability to provide a sustained treatment. Furthermore, cells, particularly stem cells, can adapt to their environment and are thus able to evolve with the pathology of the brain and spinal cord.

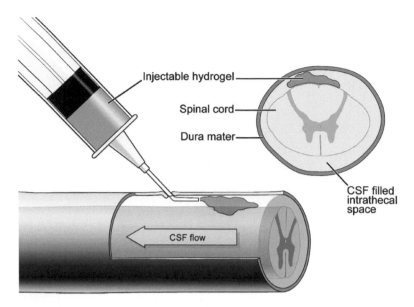

FIGURE 8.3 Injectable drug delivery to the intrathecal space. When injected into the intrathecal space, a hydrogel can localize and modulate drug release at the site of injury. This route is preferred over epidural delivery when the diffusive barrier presented by the pia mater is significant, as with large-molecular-weight therapeutics. (Copyright 2005. With permission from Michael Corrin.)

Options for the choice of cell type and source are extensive, each with distinct advantages and disadvantages. Exhaustive reviews of the different cell types and sources used experimentally following TBI and SCI were recently completed [95,96]. Stem cells are being investigated for transplantation, owing largely to their proliferative and pluri- and multipotent nature. Though transplantation of stem cells at various points along the differentiation continuum (of both neural and nonneural lineages) has been investigated, implantation of primed stem cells is a common method of current investigation because of the risk of tumorigenesis of undifferentiated stem cells and the desire to control stem cell fate. In addition to being a great source of trophic support for the rescue of host cells after CNS trauma, stem cells have the potential to directly replace the cells of the CNS, that is, neurons, astrocytes, and oligodendrocytes, all of which are damaged by the traumatic insult [97–109]. The fate of donor stem cells is dictated by both *in vitro* preparation and the host environment [110]. Using stem or progenitor cells also offers the potential for "off-the-shelf" treatments using either embryonic stem cell–derived cells or autologous grafts such as bone marrow–derived mesenchymal stem cells [111] or skin-derived stem cells [112].

Neural stem cells (NSCs) are multipotent stem cells with the capacity to differentiate into the major cells of the CNS and have many potential applications in transplantation. NSCs persist in the adult brain [56,113] and contribute to neurogenesis that occurs throughout adult mammalian life in the olfactory and hippocampal regions [56–58,113]. The rate of neuro- and gliogenesis increases following TBI [56–58,113] and is thought to be an attempt at self-repair and plasticity. Transplanting NSCs into the injured brain may augment the neuro- and gliogenic environment that the brain inherently attempts to create following injury. NSCs transplanted after experimental TBI have been shown to promote motor and cognitive recovery [96,114–117]. NSCs have also demonstrated benefit when implanted after experimental SCI [72,118]. In addition to fetal or adult sources, NSCs may be derived from embryonic- [119–122], skin- [123], or bone-marrow-derived stem cells [124,125], offering promising alternative cell sources.

Transplantation of cells derived from fetal nervous and hematopoietic tissue has already shown promise in the clinic for treating severe TBI [126]. For SCI, transplantation of embryonic stem cell–derived oligodendrocyte precursors is being pursued for clinical trials [127], where the resulting oligodendrocytes are expected to myelinate degenerated fibers, thereby providing a neuroprotective effect against degeneration. Olfactory ensheathing glia (OEG), currently in clinical trials, have been shown to guide axon regeneration and to myelinate these axons [128].

A common problem with cell transplants is their limited survival and interaction with host tissue, particularly compared to fetal tissue grafts. Transplantation of fetal tissue has been shown to promote recovery after CNS injury as well as combat neurodegenerative diseases (for reviews, see [129,130]). Studies that directly compare cells in suspension with tissue transplants have shown that donor cells in the tissue had significantly improved survival; furthermore, animals receiving intact tissue had significantly better functional outcome compared to those treated with cell suspension grafts [131–133]. The enhanced donor cell survival observed for transplantations of intact tissue may be due to the presence of three-dimensional architecture and a higher accessibility of extracellular adhesive proteins to which the donor cells can attach. Thus, tissue engineering approaches that emulate tissue transplants are being explored. While cell transplantation has already demonstrated promise in the clinic for both TBI and SCI, it is important to enhance survival and integration of donor cells in host tissue to further advance cell transplantation therapy.

Other hurdles associated with translating cell transplantation to the clinic include the host immune response to the cells or shed antigens [134] and other associated risks. The choice of cell type and source is critical in addressing these issues. For example, autologous stem cells (e.g., from the bone marrow or skin) may overcome some of these concerns. Embryonic stem cells have relatively low levels of major histocompatibility complex, thus minimizing the immune response, yet, as discussed above, priming embryonic stem cells is desirable for control toward specific cell lineages. Stimulation of endogenous stem cells could overcome hurdles associated with stem cell transplantation. In addition to determining the optimal cell type and source, the delivery time, location, and method (e.g., cell number, delivery vehicle) are important considerations. All of these factors will affect the efficacy of the treatment, and it is likely that multiple combinations of these parameters will prove to be beneficial at promoting functional recovery.

8.3.3 Tissue Engineering and Biomaterial Strategies in the Injured CNS

Tissue engineering strategies include the introduction of natural or synthetic biomaterial-based interventions as well as combinations of cells and biomaterial scaffolds. Biomaterial-based strategies include those where the biomaterial itself has some therapeutic benefit or serves as a delivery vehicle for growth factors and extracellular matrix proteins, with the goal of recruiting host cells or enhancing axonal growth. When used as a delivery vehicle for cells, biomaterials must provide a suitable microenvironment for cell survival, tissue regeneration, and host tissue integration.

Unlike polymer scaffolds that are molded into a particular shape prior to implantation, the irregularly shaped cavity resulting from traumatic injury requires a scaffold that can conform to its shape. An attractive approach is a hydrogel system injected in liquid form into the lesion cavity, which then forms a three-dimensional scaffold *in situ*, allowing for minimally invasive delivery into the lesion (as shown in Figure 8.4). The injectable hydrogel can contain free or tethered drug molecules and/or suspended cells. To this end, several thermosensitive polymeric systems such as poly(N-isopropylacrylamide) [135], agarose [136], methylcellulose [137], and poly(ethylene glycol)-poly(lactic acid)-poly(ethylene glycol) tri-block polymer [138] have been investigated. Hyaluronic acid hydrogels modified with laminin have been shown to encourage cell infiltration and angiogenesis, reduce glial scar formation, and promote neurite extension when implanted into a brain lesion [139]. In addition, injection of a hyaluronan–methylcellulose blend into the intrathecal cavity that surrounds the spinal cord tissue has been shown to promote functional recovery in experimental models of SCI [91].

Regenerative strategies often look to developmental biology as a basis for design and incorporation of specific signaling molecules that are important to cell and axonal guidance in the brain and spinal cord. Permissive scaffolds can be tailored to mimic the developing brain by promoting migration of endogenous stem cells and enhancing plasticity and redevelopment following TBI. During development, axon guidance results from a combination of attractive and repulsive, long-range and short-range cues [140], and these have been incorporated into biomimetic strategies *in vitro* [141–143] but have yet to be translated to *in vivo* preparations. Hollow fiber membranes filled with brain-derived neurotrophic factor (but not presented in a gradient) have promoted axonal regeneration *in vivo* [144].

Tissue engineering strategies often include implanting of constructs containing exogenous cells in a bioactive scaffold. Specific combinations of cells and scaffolds can be designed to meet the needs of the physiologic and pathologic system of interest. Cell-seeded polymer scaffolds have been shown to increase cell adhesion, survival, and host-implant integration in many physiological systems, including the CNS [72,145–152]. To date, several efforts have been made to utilize tissue engineering approaches in the CNS, ranging from implanting encapsulated nerve growth factor–secreting cells to treat Alzheimer's disease [153] to implanting NSCs within porous polyglycolic acid scaffolds to treat hypoxia-ischemia [148]. NSCs have also been used in combination with biomaterial scaffolds for enhanced delivery of cells following TBI [151] and enhanced regeneration following SCI [72,154].

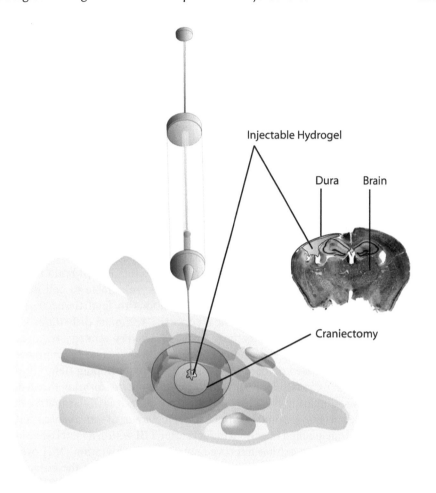

Injectable Hydrogel

Dura   Brain

Craniectomy

FIGURE 8.4 Hydrogel delivery system for traumatic brain injury. Hydrogels injected into a lesion after TBI conform to the cavity and can be rendered bioactive by tethering adhesive ligands or other molecules. When used as a drug delivery vehicle, drugs and therapeutic proteins are dissolved in, or bound to, the hydrogel. When used as a tissue engineered scaffold, the hydrogel delivers cells in the presence of signals to promote survival and control cell adhesion, migration, and differentiation.

In SCI, several studies have focused on creating a permissive environment for regeneration using either peripheral nerve grafts or biomimetic nerve guidance channels. Nerve guidance channels, or nerve cuffs, are used clinically to repair peripheral nerve injuries, providing a permissive pathway through which severed axons regrow. Some interesting results in SCI repair have been obtained using peripheral nerve grafts, where Schwann cells likely contribute by cleaning up the degenerative debris that follows injury [155–157]. Furthermore, connecting gray and white matter with a series of intercostal peripheral nerve grafts has shown promising results [158]; however, these results have been difficult to replicate. The challenge with this biomimetic approach is to stimulate a sufficient number of axons to regenerate within the defined environment. Hollow fiber membranes composed of either synthetic or naturally derived polymers have been evaluated for transected SCI models [159], and synthetic scaffolds have been tested in hemi-section models for a similar purpose [72]. Alone, the hollow fiber membranes or scaffolds are insufficient for repair, but when they are combined with regenerative factors, matrices, cells, or other drug molecules, greater regeneration has been observed. Peripheral intercostal nerves [157] and peripheral Schwann cells [160,161] have been incorporated into synthetic guidance channels or scaffolds and have demonstrated enhanced axon regeneration when combined with neurotrophic factors or hormones [155,162,163]. Another strategy is to combine the permissive channel environment with an enzyme, such as chondroitinase ABC, which has been shown to degrade the glial scar [50], thereby improving the physical pathway for regeneration [164].

In addition to using biomaterials and tissue engineering approaches to guide axon growth in the spinal cord, these strategies can be employed to improve survival and integration of cells transplanted into the traumatically injured brain or spinal cord. As noted previously, fetal tissue grafted into the injured brain has shown promising results, likely because the grafts are less vulnerable to cell death and more effectively promote repair. However, limitations to fetal tissue transplants include inadequate availability and ethical concerns, technical difficulties keeping tissues viable *in vitro*, and a potentially invasive delivery strategy required of a three-dimensional implant. These limitations can be overcome by engineering a tissue-like construct based on core components of the developing (fetal) brain tissue, such as NSCs, extracellular matrix proteins, and *in situ*–forming three-dimensional structures. Tethering bioactive ligands, such as extracellular matrix motifs, to scaffold materials may positively influence cell behavior of transplanted cells. For example, NSCs transplanted in animal models after TBI within a matrix-based scaffold have been shown to exhibit improved survival and migration [151] over cells transplanted in the absence of the scaffold, perhaps because of the antiapoptotic properties of matrix proteins such as laminin and fibronectin [165–167]. A tissue engineering approach to neural transplantation may combine the benefits of whole tissue grafts (e.g., physical and chemical support for transplanted cells leading to enhanced survival) and cell suspension transplantation (e.g., increased availability, improved cell migration, and integration with host tissue).

8.4   FUTURE OUTLOOK: OPPORTUNITIES AND CHALLENGES

Despite the numerous challenges that exist in repairing the injured CNS, there have been significant advances both in technological tools and the understanding of injury pathology mechanisms that permit development of new interventions. Numerous promising clinical trials are currently underway (or planned) for SCI, including the delivery of neutralizing molecules, cells (stem and olfactory ensheathing glia), and even biomaterials (i.e., polyethylene glycol [PEG]) [89]. While the scientific and medical community is excited by these clinical trials, a commitment has been made to combination strategies that include the use of cells, materials, and proteins. There is a strong sense that multiple pathways will have to be targeted for a substantive functional benefit to be realized. For example, cell delivery holds great promise, but technical difficulties are associated with their survival. The delivery of cells in suitable biomaterials may lead to greater donor cell survival and thus greater benefit. Future strategies will likely include factors that both promote neuroprotection and provide a suitable environment for regeneration.

In addition to the challenges facing treatment of acute injury, even greater challenges exist in treatment of chronic injury, where degeneration has persisted, muscle tissue has atrophied, cysts have formed, and the glial scar is well formed. While neuroprotective strategies may be appropriate for acute injury, regenerative strategies will be required for treatment of chronic injury. Some of the advances in tissue engineering and biomaterials may provide the underlying scaffold required to promote regeneration in the hostile environment that follows traumatic injury. For TBI, the goal of regeneration is not necessarily recreating previous circuitry, but rather, enhancing neuroplasticity (i.e., development of new or altered circuitry). Neuroplasticity is also important for SCI where axonal regeneration is required beyond the glial scar and to the target organs for restoration of the neuronal circuitry and functional improvement. Connecting the appropriate tracts poses an even greater challenge, and thus plasticity in the spinal cord and brain are critical. To date, regeneration along the same tracts has largely been ignored since the focus has been on the intermediate goal of encouraging sufficient axons to grow across the glial scar.

Because of the increased scale and scope of clinical CNS injuries, translation to human therapies is technically challenging, requiring significant amounts of cells and scaffold materials. Considerations such as scale-up, shelf-life of treatment, and invasiveness of transplant procedure are important when moving toward clinical treatments. Despite the many challenges, the future has never been so promising, with more clinical trials planned for traumatic CNS injuries than ever before and new developments in research laboratories paving the way for future combination strategies.

ACKNOWLEDGMENTS

We acknowledge contributions from the following: Jordan Wosnick, Sarah Stabenfeldt, Crystal Simon, Matthew Tate, and Brock Wester. MSS is grateful to the following agencies for funding: Canadian Institutes of Health Research, Natural Sciences and Engineering Research Council; Canada Research Chairs; and Canada Foundation for Innovation. MCL acknowledges contributions from the following agencies for funding: National Science Foundation and National Institutes of Health.

REFERENCES

1. Adelson, P. D., Dixon, C. E., and Kochanek, P. M. Long-term dysfunction following diffuse traumatic brain injury in the immature rat. *J. Neurotrauma* 17(4), 273–82, 2000.
2. Bramlett, H. M. and Dietrich, W. D. Quantitative structural changes in white and gray matter 1 year following traumatic brain injury in rats. *Acta Neuropathol.* (Berlin) 103(6), 607–14, 2002.

3. Bramlett, H. M., Green, E. J., and Dietrich, W. D. Hippocampally dependent and independent chronic spatial navigational deficits following parasagittal fluid percussion brain injury in the rat. *Brain Res.* 762(1-2), 195–202, 1997.

4. Chen, Z. L., Indyk, J. A., and Strickland, S. The hippocampal laminin matrix is dynamic and critical for neuronal survival. *Mol. Biol. Cell* 14(7), 2665–76, 2003.

5. Hall, E. D. et al. Spatial and temporal characteristics of neurodegeneration after controlled cortical impact in mice: more than a focal brain injury. *J. Neurotrauma* 22(2), 252–65, 2005.

6. Levin, H. S. et al. Neurobehavioral outcome 1 year after severe head injury. Experience of the Traumatic Coma Data Bank. *J. Neurosurg.* 73(5), 699–709, 1990.

7. Povlishock, J. T. and Katz, D. I. Update of neuropathology and neurological recovery after traumatic brain injury. *J. Head Trauma Rehabil.* 20(1), 76–94, 2005.

8. Raghupathi, R. Cell death mechanisms following traumatic brain injury. *Brain Pathol.* 14(2), 215–22, 2004.

9. Smith, D. H. et al. Progressive atrophy and neuron death for one year following brain trauma in the rat. *J. Neurotrauma* 14(10), 715–27, 1997.

10. Smith, D. H. et al. Persistent memory dysfunction is associated with bilateral hippocampal damage following experimental brain injury. *Neurosci. Lett.* 168, 151–154, 1994.

11. Whalen, M .J. et al. Reduction of cognitive and motor deficits after traumatic brain injury in mice deficient in poly(ADP-ribose) polymerase. *J. Cereb. Blood Flow Metab.* 19(8), 835–42, 1999.

12. Adekoya, N. et al. Surveillance for traumatic brain injury deaths—United States, 1989-1998. *MMWR Surveill. Summ.* 51(10), 1–14, 2002.

13. Thurman, D. and Guerrero, J. Trends in hospitalization associated with traumatic brain injury. *JAMA* 282(10), 954–57, 1999.

14. Stalhammar, D. and Starmark, J. E. Assessment of responsiveness in head injury patients. The Glasgow Coma Scale and some comments on alternative methods. *Acta Neurochir. Suppl.* (Wien) 36, 91–94, 1986.

15. Clausen, T. and Bullock, R. Medical treatment and neuroprotection in traumatic brain injury. *Curr. Pharm. Des..* 7(15), 1517–32, 2001.

16. Faden, A. I. Pharmacological treatment of central nervous system trauma. *Pharmacol. Toxicol.* 78(1), 12–17, 1996.

17. Maas, A. I. et al. Clinical trials in traumatic brain injury: current problems and future solutions. *Acta Neurochir. Suppl.* 89, 113–18, 2004.

18. Marklund, N. et al. Evaluation of pharmacological treatment strategies in traumatic brain injury. *Curr. Pharm. Des.* 12(13), 1645–80, 2006.

19. Royo, N. C. et al. Pharmacology of traumatic brain injury. *Curr. Opin. Pharmacol.* 3(1), 27–32, 2003.

20. Marmarou, A. The pathophysiology of brain edema and elevated intracranial pressure. *Cleve. Clin. J. Med.* 71(Suppl 1), S6–S8, 2004.

21. Ackery, A., Tator, C., and Krassioukov, A. A global perspective on spinal cord injury epidemiology. *J. Neurotrauma.* 21(10), 1355–70, 2004.

22. Langlios, J., Rutland-Brown, W., Thomas, K. E. *Traumatic Brain Injury in the United States: Emergency Department Visits, Hospitalizations, and Deaths.* Centers for Disease Control and Prevention, National Center for Injury Prevention and Control, Atlanta, GA, 2004.

23. Thurman, D. J. et al. Traumatic brain injury in the United States: A public health perspective. *J. Head Trauma Rehabil.* 14(6), 602–15, 1999.

24. Carlson, G. D. and Gorden, C. Current developments in spinal cord injury research. *Spine J.* 2(2), 116–28, 2002.
25. Corso, P. et al. Incidence and lifetime costs of injuries in the United States. *Inj. Prev.* 12(4), 212–18, 2006.
26. Gaetz, M. The neurophysiology of brain injury. *Clin. Neurophysiol.* 115(1), 4–18, 2004.
27. Verma, A. Opportunities for neuroprotection in traumatic brain injury. *J. Head Trauma Rehabil.* 15(5), 1149–61, 2000.
28. Fawcett, J. W. Overcoming inhibition in the damaged spinal cord. *J. Neurotrauma* 23(3-4), 371–83, 2006.
29. Correale, J. and Villa, A. The neuroprotective role of inflammation in nervous system injuries. *J. Neurol.* 251(11), 1304–16, 2004.
30. Lenzlinger, P. M. et al. The duality of the inflammatory response to traumatic brain injury. *Mol. Neurobiol.* 24(1-3), 169–81, 2001.
31. Morganti-Kossmann, M. C. et al. Inflammatory response in acute traumatic brain injury: a double-edged sword. *Curr. Opin. Crit. Care* 8(2), 101–5, 2002.
32. Schwartz, M. Autoimmune involvement in CNS trauma is beneficial if well controlled. *Prog. Brain Res.* 128, 259–63, 2000.
33. Schmidt, O. I. et al. Closed head injury—An inflammatory disease? *Brain Res. Brain Res. Rev.* 48(2), 388–99, 2005.
34. Kato, H. and Walz, W. The initiation of the microglial response. *Brain Pathol.* 10(1), 137–43, 2000.
35. Ladeby, R. et al. Microglial cell population dynamics in the injured adult central nervous system. *Brain Res. Brain Res. Rev.*. 48(2), 196–206, 2005.
36. Hauwel, M. et al. Innate (inherent) control of brain infection, brain inflammation and brain repair: The role of microglia, astrocytes, "protective" glial stem cells and stromal ependymal cells. *Brain Res. Brain Res. Rev.*. 48(2), 220–33, 2005.
37. Schwartz, M. and Yoles, E. Macrophages and dendritic cells treatment of spinal cord injury: from the bench to the clinic. *Acta Neurochir. Suppl.* 93,147–50, 2005.
38. Bao, F. et al. Early anti-inflammatory treatment reduces lipid peroxidation and protein nitration after spinal cord injury in rats. *J. Neurochem.* 88(6), 1335–44, 2004.
39. Gomes-Leal, W. et al. Astrocytosis, microglia activation, oligodendrocyte degeneration, and pyknosis following acute spinal cord injury. *Exp. Neurol.* 190(2), 456–67, 2004.
40. Mabon, P. J., Weaver, L. C., and Dekaban, G. A. Inhibition of monocyte/macrophage migration to a spinal cord injury site by an antibody to the integrin alphaD: a potential new anti-inflammatory treatment. *Exp. Neurol.* 166(1), 52–64, 2000.
41. Fouad, K. et al. Combining Schwann cell bridges and olfactory-ensheathing glia grafts with chondroitinase promotes locomotor recovery after complete transection of the spinal cord. *J. Neurosci.* 25(5), 1169–78, 2005.
42. Oudega, M. and Xu, X. M. Schwann cell transplantation for repair of the adult spinal cord. *J. Neurotrauma* 23(3-4), 453–67, 2006.
43. Eddleston, M. and Mucke, L. Molecular profile of reactive astrocytes—implications for their role in neurologic disease. *Neuroscience* 54(1), 15–36, 1993.
44. McKerracher, L. et al. Identification of myelin-associated glycoprotein as a major myelin-derived inhibitor of neurite growth. *Neuron* 13(4), 805–11, 1994.
45. Mukhopadhyay, G. et al. A novel role for myelin-associated glycoprotein as an inhibitor of axonal regeneration. *Neuron* 13(3), 757–67, 1994.
46. Fawcett, J. W. and Asher, R. A. The glial scar and central nervous system repair. *Brain Res. Bull.*. 49(6), 377–91, 1999.

47. Bandtlow, C. E. and Schwab, M. E. NI-35/250/nogo-a: a neurite growth inhibitor restricting structural plasticity and regeneration of nerve fibers in the adult vertebrate CNS. *Glia* 29(2), 175–81, 2000.
48. Properzi, F., Asher, R. A., and Fawcett, J.W. Chondroitin sulphate proteoglycans in the central nervous system: Changes and synthesis after injury. *Biochem. Soc. Trans.* 31(2), 335–36, 2003.
49. Schwab, M. E. and Bartholdi, D. Degeneration and regeneration of axons in the lesioned spinal cord. *Physiol. Rev.* 76(2), 319–70, 1996.
50. Rhodes, K. E. and Fawcett, J. W. Chondroitin sulphate proteoglycans: preventing plasticity or protecting the CNS? *J. Anat.* 204(1), 33–48, 2004.
51. Fournier, A. E., Takizawa, B. T., and Strittmatter, S. M. Rho kinase inhibition enhances axonal regeneration in the injured CNS. *J. Neurosci.* 23(4), 1416–23, 2003.
52. McKerracher, L. and Higuchi, H. Targeting Rho to stimulate repair after spinal cord injury. *J. Neurotrauma.* 23(3-4), 309–17, 2006.
53. von Meyenburg, J. et al. Regeneration and sprouting of chronically injured corticospinal tract fibers in adult rats promoted by NT-3 and the mAb IN-1, which neutralizes myelin-associated neurite growth inhibitors. *Exp. Neurol.* 154(2), 583–94, 1998.
54. Tate, C. C., Tate, M. C., and LaPlaca, M. C. Fibronectin and laminin increase in the mouse brain after controlled cortical impact injury. *J. Neurotrauma* 24(1), 226–30, 2007.
55. David, S. and Aguayo, A. J. Axonal elongation into peripheral nervous system "bridges" after central nervous system injury in adult rats. *Science* 214(4523), 931–33, 1981.
56. Dash, P. K., Mach, S. A., and Moore, A. N. Enhanced neurogenesis in the rodent hippocampus following traumatic brain injury. *J. Neurosci. Res.* 63(4), 313–19, 2001.
57. Kernie, S. G., Erwin, T. M., and Parada, L. F. Brain remodeling due to neuronal and astrocytic proliferation after controlled cortical injury in mice. *J. Neurosci. Res.* 66(3), 317–26, 2001.
58. Ramaswamy, S. et al. Cellular proliferation and migration following a controlled cortical impact in the mouse. *Brain Res.* 1053(1-2), 38–53, 2005.
59. Fehlings, M. G. and Perrin, R. G. The role and timing of early decompression for cervical spinal cord injury: Update with a review of recent clinical evidence. *Injury* 36(Suppl 2), B13–26, 2005.
60. Sahuquillo, J. and Arikan, F. Decompressive craniectomy for the treatment of refractory high intracranial pressure in traumatic brain injury. *Cochrane Database Syst. Rev.* (1), CD003983, 2006.
61. Roberts, I., Schierhout, G., and Alderson, P. Absence of evidence for the effectiveness of five interventions routinely used in the intensive care management of severe head injury: a systematic review [see comments]. *J. Neurol. Neurosurg. Psychiatry* 65(5), 729–33, 1998.
62. Ikonomidou, C. and Turski, L. Why did NMDA receptor antagonists fail clinical trials for stroke and traumatic brain injury? *Lancet Neurol.* 1(6), 383–86, 2002.
63. Willis, C., Lybrand, S., and Bellamy, N. Excitatory amino acid inhibitors for traumatic brain injury. *Cochrane Database Syst. Rev.* (1), CD003986, 2004.
64. Narayan, R. K. et al. Clinical trials in head injury. *J. Neurotrauma* 19(5), 503–57, 2002.
65. Sahuquillo, J. , et al. Moderate hypothermia in the management of severe traumatic brain injury: A good idea proved ineffective? *Curr. Pharm. Des.* 10(18), 2193–204, 2004.

66. Cutler, S. M. et al. Slow-release and injected progesterone treatments enhance acute recovery after traumatic brain injury. *Pharmacol. Biochem. Behav.* 84(3), 420–28, 2006.

67. Stein, D. G. and Fulop, Z. L. Progesterone and recovery after traumatic brain injury: an overview. *Neuroscientist* 4(6), 435–42, 1998.

68. Wang, K. K. et al. Neuroprotection targets after traumatic brain injury. *Curr. Opin. Neurol.* 19(6), 514–19, 2006.

69. Bracken, M. B. et al. Methylprednisolone or naloxone treatment after acute spinal cord injury: 1-year follow-up data. Results of the second National Acute Spinal Cord Injury Study. *J. Neurosurg.* 76(1), 23–31, 1992.

70. Bracken, M. B. et al. Methylprednisolone or tirilazad mesylate administration after acute spinal cord injury: 1-year follow up. Results of the third National Acute Spinal Cord Injury randomized controlled trial. *J. Neurosurg.* 89(5), 699–706, 1998.

71. Ramer, L. M., Ramer, M. S., and Steeves, J. D. Setting the stage for functional repair of spinal cord injuries: A cast of thousands. *Spinal Cord* 43(3), 134–61, 2005.

72. Teng, Y. D. et al. Functional recovery following traumatic spinal cord injury mediated by a unique polymer scaffold seeded with neural stem cells. *Proc. Natl. Acad. Sci. USA* 99(5), 3024–29, 2002.

73. Young, W. Recovery mechanisms in spinal cord injury: implications for regenerative therapy. In *Neural Regeneration and Transplantation*, Vol. 6 of *Frontiers of Clinical Neuroscience*, F. J. Seil and A. R. Liss, Eds. Williams & Wilkins, New York, 1989, pp. 157–69.

74. Wells, J. E. et al. Neuroprotection by minocycline facilitates significant recovery from spinal cord injury in mice. *Brain* 126(Pt 7), 1628–37, 2003.

75. Pearse, D. D. et al. cAMP and Schwann cells promote axonal growth and functional recovery after spinal cord injury. *Nat. Med.* 10(6), 610–16, 2004.

76. Geisler, F. H. et al. The Sygen multicenter acute spinal cord injury study. *Spine* 26(24 Suppl), S87–98, 2001.

77. Olson, L. et al. Intraputaminal infusion of nerve growth factor to support adrenal medullary autografts in Parkinson's disease. One year follow-up of first clinical trial. *Arch. Neurol.* 48, 373–81, 1991.

78. Olson, L. et al. Nerve growth factor affects 11C-nicotine binding, blood flow, EEG, and verbal episodic memory in an Alzheimer patient. *J. Neural. Transm. Park. Dis. Dement. Sect.* 4, 79–95, 1992.

79. Saltzman, W. M. et al. Intracranial delivery of recombinant nerve growth factor: release kinetics and protein distribution for three delivery systems. *Pharm. Res.* 16(2), 232–40, 1999.

80. Aebischer, P. et al. Transplantation of polymer encapsulated neurotransmitter secreting cells: effect of the encapsulation technique. *J. Biomech. Eng.* 113(2), 178–83, 1991.

81. During, M. J. et al. Biochemical and behavioral recovery in a rodent model of Parkinson's disease following stereotactic implantation of dopamine-containing liposomes. *Exp. Neurol.* 115(2), 193–99, 1992.

82. Tabata, Y., Gutta, S., and Langer, R. Controlled delivery systems for proteins using polyanhydride microspheres. *Pharm. Res.* 10(4), 487–96, 1993.

83. Krewson, C. E., Klarman, M. L., and Saltzman, W. M. Distribution of nerve growth factor following direct delivery to brain interstitium. *Brain Res.* 680(1-2), 196–206, 1995.

84. Amar, A. P., Larsen, D. W., and Teitelbaum, G. P. Percutaneous spinal interventions. *Neurosurg. Clin. N. Am.* 16(3), 561–68, 2005.

85. Paavola, A. et al. Controlled release gel of ibuprofen and lidocaine in epidural use—analgesia and systemic absorption in pigs. *Pharm. Res.* 15(3), 482–78, 1998.
86. Paavola, A. et al. Controlled release injectable liposomal gel of ibuprofen for epidural analgesia. *Int. J. Pharm.* 199(1), 85–93, 2000.
87. Dergham, P. et al. Rho signaling pathway targeted to promote spinal cord repair. *J. Neurosci.* 22(15), 6570–77, 2002.
88. Ethans, K. D. et al. Intrathecal drug therapy using the Codman Model 3000 Constant Flow Implantable Infusion Pumps: experience with 17 cases. *Spinal Cord* 43(4), 214–18, 2005.
89. Tator, C. H. Review of treatment trials in human spinal cord injury: issues, difficulties, and recommendations. *Neurosurgery* 59(5), 957–82; discussion 982–87, 2006.
90. Jones, L. L. and Tuszynski, M. H. Chronic intrathecal infusions after spinal cord injury cause scarring and compression. *Microsc. Res. Tech.* 54(5), 317–324, 2001.
91. Gupta, D., Tator, C. H., and Shoichet, M. S. Fast-gelling injectable blend of hyaluronan and methylcellulose for intrathecal, localized delivery to the injured spinal cord. *Biomaterials* 27(11), 2370–79, 2006.
92. Jimenez Hamann, M. C., Tator, C. H., and Shoichet, M. S. Injectable intrathecal delivery system for localized administration of EGF and FGF-2 to the injured rat spinal cord. *Exp. Neurol.* 194(1), 106–19, 2005.
93. Jimenez Hamann, M. C. et al. Novel intrathecal delivery system for treatment of spinal cord injury. *Exp. Neurol.* 182(2), 300–309, 2003.
94. Kleindienst, A. et al. Enhanced hippocampal neurogenesis by intraventricular S100B infusion is associated with improved cognitive recovery after traumatic brain injury. *J. Neurotrauma.* 22(6), 645–55, 2005.
95. Enzmann, G. U. et al. Functional considerations of stem cell transplantation therapy for spinal cord repair. *J. Neurotrauma* 23(3-4), 479–95, 2006.
96. Longhi, L. et al. Stem cell transplantation as a therapeutic strategy for traumatic brain injury. *Transpl. Immunol.* 15(2), 143–48, 2005.
97. Chopp, M., Li, Y., and Jiang, N. Increase in apoptosis and concomitant reduction of ischemic lesion volume and evidence for synaptogenesis after transient focal cerebral ischemia in rat treated with staurosporine. *Brain Res.* 828(1-2), 197–201, 1999, 1998.
Kondziolka, D., Wechsler, L., and Achim, C. Neural transplantation for stroke. *J. Clin. Neurosci.* 9(3), 225–30, 2002.
99. Liu, Q. et al. Preparation of macroporous poly(2-hydroxyethyl methacrylate) hydrogels by enhanced phase separation. *Biomaterials* 21(21), 2163–69, 2000.
100. Llado, J. et al. Neural stem cells protect against glutamate-induced excitotoxicity and promote survival of injured motor neurons through the secretion of neurotrophic factors. *Mol. Cell. Neurosci.* 27(3), 322–31, 2004.
101. Lu, P. et al. Neural stem cells constitutively secrete neurotrophic factors and promote extensive host axonal growth after spinal cord injury. *Exp. Neurol.* 181(2), 115–29, 2003.
102. Mahmood, A. et al. Long-term recovery after bone marrow stromal cell treatment of traumatic brain injury in rats. *J. Neurosurg.* 104(2), 272–77, 2006.
103. Ourednik, J. et al. Neural stem cells display an inherent mechanism for rescuing dysfunctional neurons. *Nat. Biotechnol.* 20(11), 1103–10, 2002.
104. Pluchino, S. et al. Neurosphere-derived multipotent precursors promote neuroprotection by an immunomodulatory mechanism. *Nature* 436(7048), 266–71, 2005.
105. Rafuse, V. F. et al. Neuroprotective properties of cultured neural progenitor cells are associated with the production of sonic hedgehog. *Neuroscience* 131(4), 899–916, 2005.

106. Royo, N. C. et al. From cell death to neuronal regeneration: building a new brain after traumatic brain injury. *J. Neuropathol. Exp. Neurol.* 62(8), 801–11, 2003.

107. Sladek, J. R., Jr., Redmond, D. E., Jr., and Roth, R. H. Transplantation of fetal neurons in primates. *Clin. Res.*. 36(3), 200–204, 1988.

108. Wong, A. M. Hodges, H., and Horsburgh, K. Neural stem cell grafts reduce the extent of neuronal damage in a mouse model of global ischaemia. *Brain Res.* 1063(2), 140–50, 2005.

109. Yan, J. et al. Differentiation and tropic/trophic effects of exogenous neural precursors in the adult spinal cord. *J. Comp. Neurol.* 480(1), 101–14, 2004.

110. Yandava, B. D., Billinghurst, L.L., and Snyder, E.Y. "Global" cell replacement is feasible via neural stem cell transplantation: Evidence from the dysmyelinated shiverer mouse brain. *Proc. Natl. Acad. Sci. USA* 96(12), 7029–34, 1999.

111. Gao, J. et al. Tissue-engineered fabrication of an osteochondral composite graft using rat bone marrow-derived mesenchymal stem cells. *Tissue Eng.* 7(4), 363–71, 2001.

112. Toma, J. G. et al. Isolation of multipotent adult stem cells from the dermis of mammalian skin. *Nat. Cell Biol.* 3(9), 778–84, 2001.

113. Johansson, C. B. et al. Neural stem cells in the adult human brain. *Exp. Cell Res.* 253(2), 733–36, 1999.

114. Philips, M. F. et al. Neuroprotective and behavioral efficacy of nerve growth factor-transfected hippocampal progenitor cell transplants after experimental traumatic brain injury. *J. Neurosurg.* 94(5), 765–74, 2001.

115. Riess, P. et al. Transplanted neural stem cells survive, differentiate, and improve neurological motor function after experimental traumatic brain injury. *Neurosurgery* 51(4), 1043–52; discussion 1052–54, 2002.

116. Schouten, J. W. et al. A review and rationale for the use of cellular transplantation as a therapeutic strategy for traumatic brain injury. *J. Neurotrauma* 21(11), 1501–38, 2004.

117. Shear, D. A. et al. Neural progenitor cell transplants promote long-term functional recovery after traumatic brain injury. *Brain Res.* 1026(1), 11–22, 2004.

118. Karimi-Abdolrezaee, S. et al. Delayed transplantation of adult neural precursor cells promotes remyelination and functional neurological recovery after spinal cord injury. *J. Neurosci.* 26(13), 3377–89, 2006.

119. Brustle, O. et al. In vitro-generated neural precursors participate in mammalian brain development. *Proc. Natl. Acad. Sci. USA* 94(26), 14809–14, 1997.

120. Okabe, S. et al. Development of neuronal precursor cells and functional postmitotic neurons from embryonic stem cells *in vitro. Mech. Dev.* 59(1), 89–102, 1996.

121. Reubinoff, B.E. et al. Neural progenitors from human embryonic stem cells. *Nat. Biotechnol.* 19(12), 1134–40, 2001.

122. Tropepe, V. et al. Direct neural fate specification from embryonic stem cells: a primitive mammalian neural stem cell stage acquired through a default mechanism. *Neuron* 30(1), 65–78, 2001.

123. McKenzie, I. A. et al. Skin-derived precursors generate myelinating Schwann cells for the injured and dysmyelinated nervous system. *J. Neurosci.* 26(24), 6651–60, 2006.

124. Deng, W. et al. *In vitro* differentiation of human marrow stromal cells into early progenitors of neural cells by conditions that increase intracellular cyclic AMP. *Biochem. Biophys. Res. Commun.* 282(1), 148–52, 2001.

125. Kabos, P. et al. Generation of neural progenitor cells from whole adult bone marrow. *Exp. Neurol.* 178(2), 288–93, 2002.

126. Seledtsov, V. I. et al. Cell transplantation therapy in re-animating severely head-injured patients. *Biomed. Pharmacother.* 59(7), 415–20, 2005.

127. Faulkner, J. and Keirstead, H. S. Human embryonic stem cell-derived oligodendrocyte progenitors for the treatment of spinal cord injury. *Transpl. Immunol.* 15(2), 131–42, 2005.

128. Ramon-Cueto, A. et al. Long-distance axonal regeneration in the transected adult rat spinal cord is promoted by olfactory ensheathing glia transplants. *J. Neurosci.* 18(10), 3803–15, 1998.

129. Koutouzis, T. K. et al. Cell transplantation for central nervous system disorders. *Crit. Rev. Neurobiol.* 8(3), 125–62, 1994.

130. Subramanian, T. Cell transplantation for the treatment of Parkinson's disease. *Semin. Neurol.* 21(1), 103–15, 2001.

131. Clarkson, E. D. et al. Strands of embryonic mesencephalic tissue show greater dopamine neuron survival and better behavioral improvement than cell suspensions after transplantation in Parkinsonian rats. *Brain Res.* 806(1), 60–68, 1998.

132. Hoovler, D. W. and Wrathall, J. R. Implantation of neuronal suspensions into contusive injury sites in the adult rat spinal cord. *Acta Neuropathol.* 81(3), 303–11, 1991.

133. Sinson, G., Voddi, M., and McIntosh, T. K. Combined fetal neural transplantation and nerve growth factor infusion: Effects on neurological outcome following fluid-percussion brain injury in the rat. *J. Neurosurg.* 84(4), 655–62, 1996.

134. Jones, K. S., Sefton, M. V., and Gorczynski, R. M. *In vivo* recognition by the host adaptive immune system of microencapsulated xenogeneic cells. *Transplantation* 78(10), 1454–62, 2004.

135. Stile, R. A. et al. Poly(N-isopropylacrylamide)-based semi-interpenetrating polymer networks for tissue engineering applications. Effects of linear poly(acrylic acid) chains on rheology. *J. Biomater. Sci. Polym. Ed.* 15(7), 865–78, 2004.

136. Yu, X. and Bellamkonda, R. V. Tissue-engineered scaffolds are effective alternatives to autografts for bridging peripheral nerve gaps. *Tissue Eng.* 9(3), 421–30, 2003.

137. Tate, M. C. et al. Biocompatibility of methylcellulose-based constructs designed for intracerebral gelation following experimental traumatic brain injury. *Biomaterials* 22(10), 1113–23, 2001.

138. Jeong, B., Kim, S. W., and Bae, Y. H. Thermosensitive sol-gel reversible hydrogels. *Adv. Drug Deliv. Rev.* 54(1), 37–51, 2002.

139. Hou, S. et al. The repair of brain lesion by implantation of hyaluronic acid hydrogels modified with laminin. *J. Neurosci. Methods* 148(1), 60–70, 2005.

140. Tessier-Lavigne, M. and Goodman, C. S. The molecular biology of axon guidance. *Science* 274(5290), 1123–33, 1996.

141. Cao, X. and Shoichet, M. S. Investigating the synergistic effect of combined neurotrophic factor concentration gradients to guide axonal growth. *Neuroscience* 122(2), 381–89, 2003.

142. Kapur, T. A. and Shoichet, M. S. Immobilized concentration gradients of nerve growth factor guide neurite outgrowth. *J. Biomed. Mater. Res. A* 68(2), 235–43, 2004.

143. Moore, K., MacSween, M. and Shoichet, M. Immobilized concentration gradients of neurotrophic factors guide neurite outgrowth of primary neurons in macroporous scaffolds. *Tissue Eng.* 12(2), 267–78, 2006.

144. Patist, C. M. et al. Freeze-dried poly(D,L-lactic acid) macroporous guidance scaffolds impregnated with brain-derived neurotrophic factor in the transected adult rat thoracic spinal cord. *Biomaterials* 25(9), 1569–82, 2004.

145. Dai, W., Belt, J., and Saltzman, W. M. Cell-binding peptides conjugated to poly(ethylene glycol) promote neural cell aggregation. *Biotechnology* (NY) 12(8), 797–801, 1994.

146. Hern, D. L. and Hubbell, J. A. Incorporation of adhesion peptides into nonadhesive hydrogels useful for tissue resurfacing. *J. Biomed. Mater. Res.* 39(2), 266–76, 1998.

147. Mann, B. K., Schmedlen, R. H., and West, J. L. Tethered-TGF-beta increases extracellular matrix production of vascular smooth muscle cells. *Biomaterials* 22(5), 439–44, 2001.

148. Park, K. I., Teng, Y. D., and Snyder, E. Y. The injured brain interacts reciprocally with neural stem cells supported by scaffolds to reconstitute lost tissue. *Nat. Biotechnol.* 20(11), 1111–17, 2002.

149. Silva, G. A. et al. Selective differentiation of neural progenitor cells by high-epitope density nanofibers. *Science* 303(5662), 1352–55, 2004.

150. Stabenfeldt, S. E., Garcia, A. J., and Laplaca, M. C. Thermoreversible laminin-functionalized hydrogel for neural tissue engineering. *J. Biomed. Mater. Res. A* 77(4), 718–25, 2006.

151. Tate, M. C. et al. Fibronectin promotes survival and migration of primary neural stem cells transplanted into the traumatically injured mouse brain. *Cell Transplant.* 11(3), 283–95, 2002.

152. Yu, X., Dillon, G. P., and Bellamkonda, R. B. A laminin and nerve growth factor-laden three-dimensional scaffold for enhanced neurite extension. *Tissue Eng.* 5(4), 291–304, 1999.

153. Winn, S. R. et al. Polymer-encapsulated genetically modified cells continue to secrete human nerve growth factor for over one year in rat ventricles: behavioral and anatomical consequences. *Exp. Neurol.* 140(2), 126–38, 1996.

154. Lavik, E. et al. Seeding neural stem cells on scaffolds of PGA, PLA, and their copolymers. *Methods Mol. Biol.* 198, 89–97, 2002.

155. Bunge, M. B. Bridging the transected or contused adult rat spinal cord with Schwann cell and olfactory ensheathing glia transplants. *Prog. Brain Res.* 137, 275–82, 2002.

156. Nomura, H., Tator, C. H., and Shoichet, M. S. Bioengineered strategies for spinal cord repair. *J. Neurotrauma* 23(3-4), 496–507, 2006.

157. Tsai, E. C., Krassioukov, A. V., and Tator, C. H. Corticospinal regeneration into lumbar grey matter correlates with locomotor recovery after complete spinal cord transection and repair with peripheral nerve grafts, fibroblast growth factor 1, fibrin glue, and spinal fusion. *J. Neuropathol. Exp. Neurol.* 64(3), 230–44, 2005.

158. Cheng, H., Cao, Y., and Olson, L. Spinal cord repair in adult paraplegic rats: partial restoration of hind limb function. *Science* 273(5274), 510–13, 1996.

159. Tsai, E. C. et al. Synthetic hydrogel guidance channels facilitate regeneration of adult rat brainstem motor axons after complete spinal cord transection. *J. Neurotrauma* 21(6), 789–804, 2004.

160. Novikov, L. N. et al. A novel biodegradable implant for neuronal rescue and regeneration after spinal cord injury. *Biomaterials* 23(16), 3369–76, 2002.

161. Oudega, M. et al. Axonal regeneration into Schwann cell grafts within resorbable poly(alpha-hydroxyacid) guidance channels in the adult rat spinal cord. *Biomaterials* 22(10), 1125–36, 2001.

162. Jones, L. L. et al. Neurotrophic factors, cellular bridges and gene therapy for spinal cord injury. *J. Physiol.* 533(Pt 1), 83–89, 2001.

163. Kamada, T. et al. Transplantation of bone marrow stromal cell-derived Schwann cells promotes axonal regeneration and functional recovery after complete transection of adult rat spinal cord. *J. Neuropathol. Exp. Neurol.* 64(1), 37–45, 2005.

164. Friedman, J. A. et al. Biodegradable polymer grafts for surgical repair of the injured spinal cord. *Neurosurgery* 51(3), 742–51; discussion 751–52, 2002.

165. Gary, D. S. and Mattson, M. P. Integrin signaling via the PI3-kinase-Akt pathway increases neuronal resistance to glutamate-induced apoptosis. *J. Neurochem.* 76(5), 1485–96, 2001.

166. Gary, D. S. et al. Essential role for integrin linked kinase in Akt-mediated integrin survival signaling in hippocampal neurons. *J. Neurochem.* 84(4), 878–90, 2003.
167. Gu, Z. et al. A highly specific inhibitor of matrix metalloproteinase-9 rescues laminin from proteolysis and neurons from apoptosis in transient focal cerebral ischemia. *J. Neurosci.* 25(27), 6401–408, 2005.

Index

Bold locators indicate material in figures and tables.

C

Calcium
alginate gels and, 190–191
axons and, 17
in bone remodeling, 29
ionized calcium-binding adaptor molecule, 108
myelin and, 17
in PNS repair, 17
smooth muscle cells and, 28
stress and, 164
Calcium carbon trioxide, 190
Calcium chloride, 190
Calcium phosphate, 20, 190
Calcium sulfate, 190
Callus
angiogenesis and, 18
FGF and, 18
hard, **15**, 18
histological analysis of, **19**
PDGF and, 18
in remodeling stage, **26**, 28–29
in repair stage, **15**, 18
soft, **15**, 18, **19**
TGF-β and, 18
Calpains, 161
Calves, 77–78
cAMP moderators, 228
Cancer, 77, 78
Capacitance of electrodes, 49, **50**
Capillaries
in bone remodeling, 25
in bone repair, 14, 20
in CNS, 4
in cortex, 154–155
end-feet and, 4
heat and, 72, 77–78, 81
in repair stage, 12
in skin remodeling, 28
in skin repair, **11**, 18
Capillary blood perfusion coefficient, 72
Cartilage, 12, 20
Cathepsins, 161
Catheter/mini-pump system, 228, **229**
cBMI; *See* Cortex brain–machine interface
CD45, 8
CDPGYIGSR nonapeptide, 96, 180, 189; *See also* YIGSR
Cell adhesion L-1 molecule, 169
Cell lines
astrocyte, 97, **98**, **99**
BV-2, 98
CRL-1213, 95
CRL-1690, 97
CRL-2005, 97
C6, 97, **98**
description of, 94–98

DITNC1, 97
fibroblast, 95, **96**
F98, 97
HAPI, 98, **100**
LRM55, 97, **99**
microglia, 97–98
neuronal, 95–96
NG108-15, 195
PC-12; *See* Pheochromocytoma cell line
SH-SY5Y; *See* SH-SY5Y cell line
SK-N-SH, 209
3T3 fibroblast, 95, **96**
Cell phones, 75
Cell suspension grafts, 231, 234
Center for Neural Communication Technology (CNCT), 179
Central nervous system (CNS)
astrocytes in, 158
axons in; *See* Axons, CNS
blood-brain barrier; *See* Blood-brain barrier
blood vessels in, 4
brain; *See* Brain
cerebrum, 9
charge and, 51–52
dual-probe microdialysis studies of, 119, 139, 143
to ECS diffusivity; *See* CNS–ECS solute diffusion
ED1 marker for studies, 122
endothelial cells in, 4
gel diffusion in; *See* Gel diffusion *in vitro* experiments
glial cells in, 4
heat and, 78
hemostasis in, **5**, 6
"immunological privilege" in, 11
inflammation stage in; *See* Inflammation stage
IOI studies of, 119, 139, 142–143
leukocyte activity in, 8
lymphocytes in, 9–12
myelin in; *See* Myelin, CNS
remodeling stage in, **5**, 29–30
repair stage in, **5**, **10**, 12, 20–24, 226–234
schematic of injury to, **224**
spinal cord; *See* Spinal cord
TAD studies of; *See* CNS–ECS TAD studies
T-cells in, 9–12
tissue diffusivity model, 126–129
TMA+ ISM studies of, 119, 139, 143
wound healing in, **5**, **10**, **154**, **224**
Cerebellar granule neurons, 104
Cerebral glia, 102, 104
Cerebrospinal fluid (CSF), 9–12, **22**, 30, **230**
Cerebrum, 9
Cethrin, 225, 228
Charge

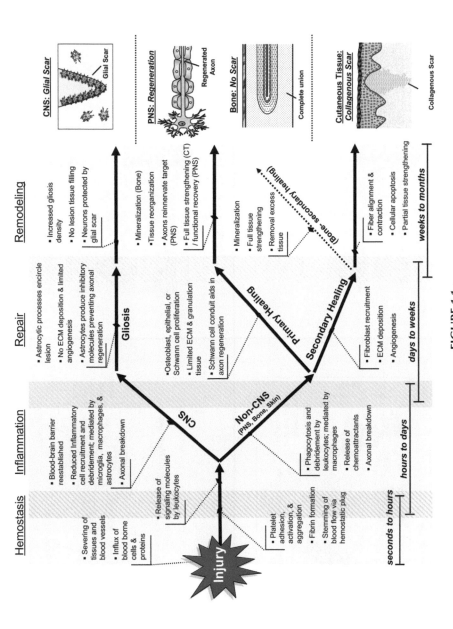

Hemostasis   **Inflammation**   **Repair**   **Remodeling**

seconds to hours   *hours to days*   *days to weeks*   *weeks to months*

Injury

- Severing of tissues and blood vessels
- Influx of blood borne cells & proteins
- Release of signaling molecules by leukocytes

- Platelet adhesion, activation, & aggregation
- Fibrin formation
- Stemming of blood flow via hemostatic plug

CNS

- Blood-brain barrier reestablished
- Reduced Inflammatory cell recruitment and debridement; mediated by microglia, macrophages, & astrocytes
- Axonal breakdown

Non-CNS (PNS, Bone, Skin)

- Phagocytosis and debridement by leukocytes; mediated by macrophages
- Release of chemoattractants
- Axonal breakdown

Gliosis

- Astrocytic processes encircle lesion
- No ECM deposition & limited angiogenesis
- Astrocytes produce inhibitory molecules preventing axonal regeneration

- Osteoblast, epithelial, or Schwann cell proliferation
- Limited ECM & granulation tissue
- Schwann cell conduit aids in axon regeneration

Primary Healing

Secondary Healing

- Fibroblast recruitment
- ECM deposition
- Angiogenesis

- Increased gliosis density
- No lesion tissue filling
- Neurons protected by glial scar

- Mineralization (Bone)
- Tissue reorganization
- Axons reinnervate target (PNS)
- Full tissue strengthening (CT) / functional recovery (PNS)

- Mineralization
- Full tissue strengthening
- Removal excess tissue

(Bone - secondary healing)

- Fiber alignment & contraction
- Cellular apoptosis
- Partial tissue strengthening

CNS: *Glial Scar*

Glial Scar

PNS: *Regeneration*

Regenerated Axon

Bone: *No Scar*

Complete union

Cutaneous Tissue: *Collagenous Scar*

Collagenous Scar

FIGURE 1.1

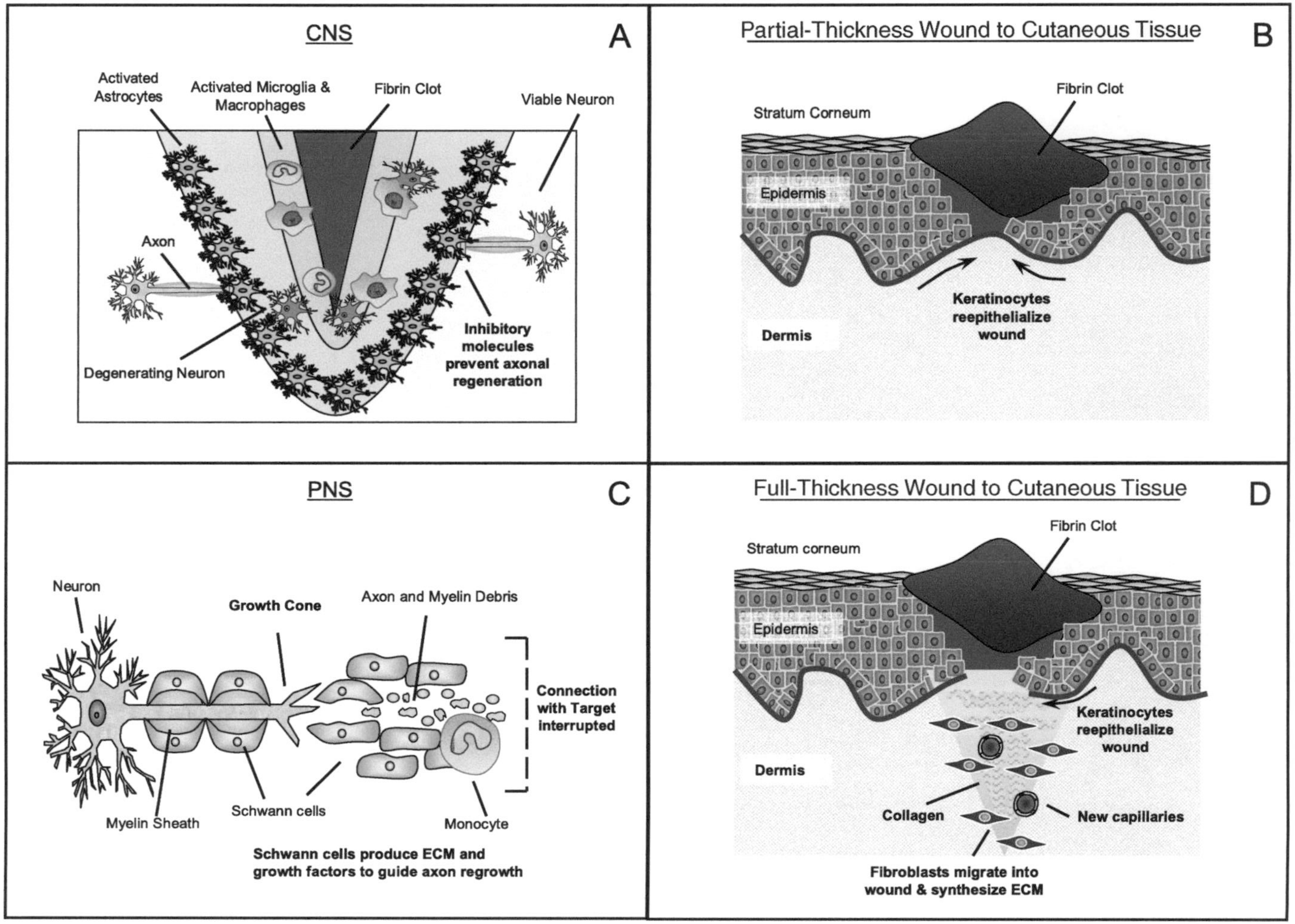

B

Partial-Thickness Wound to Cutaneous Tissue

Fibrin Clot

Stratum Corneum

Epidermis

Keratinocytes reepithelialize wound

Dermis

D

Full-Thickness Wound to Cutaneous Tissue

Fibrin Clot

Stratum corneum

Epidermis

Keratinocytes reepithelialize wound

New capillaries

Collagen

Fibroblasts migrate into wound & synthesize ECM

Dermis

A

CNS

Viable Neuron

Inhibitory molecules prevent axonal regeneration

Fibrin Clot

Activated Microglia & Macrophages

Activated Astrocytes

Axon

Degenerating Neuron

C

PNS

Connection with Target interrupted

Axon and Myelin Debris

Monocyte

Growth Cone

Schwann cells

Neuron

Myelin Sheath

Schwann cells produce ECM and growth factors to guide axon regrowth

FIGURE 1.2

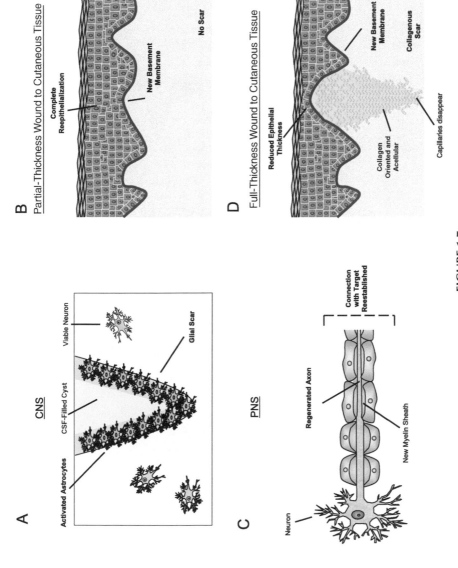

FIGURE 1.7

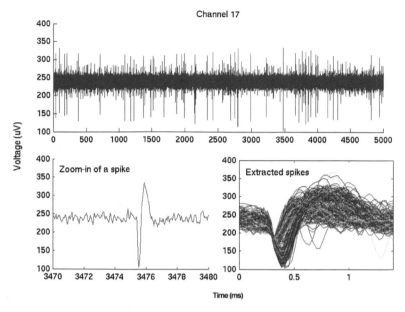

FIGURE 3.2

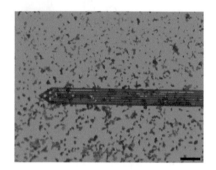

FIGURE 4.2

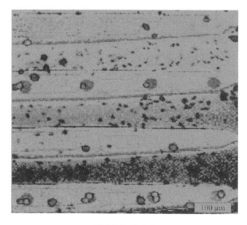

FIGURE 4.3

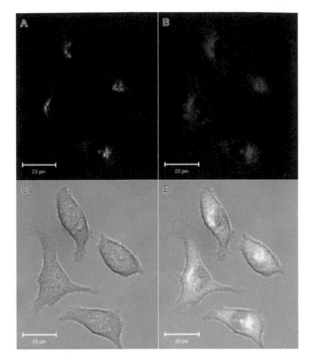

FIGURE 4.5

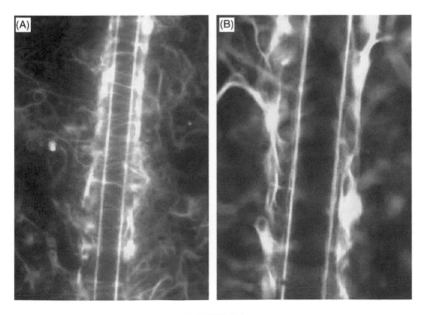

FIGURE 4.6

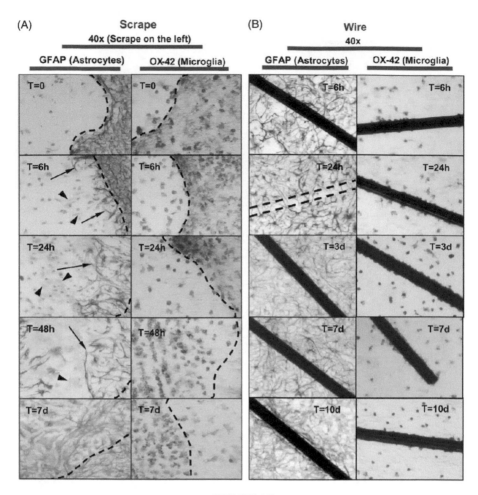

FIGURE 4.7

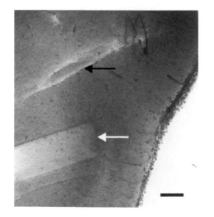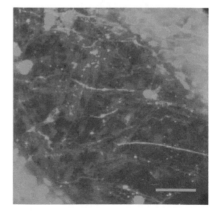

FIGURE 4.9

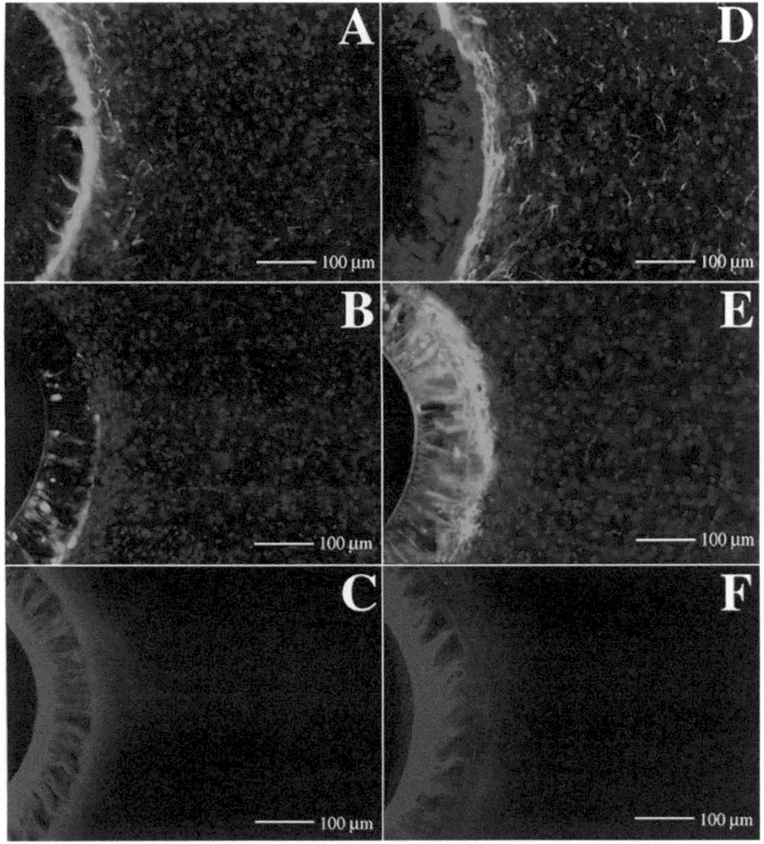

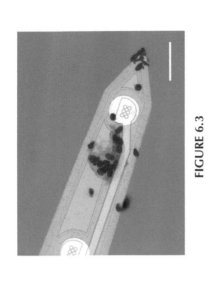

A
D
B
E
C
F
100 μm
100 μm
100 μm
100 μm
100 μm
100 μm

FIGURE 5.6

FIGURE 6.3

.1
.01
.100E-02
.100E-03
a
b
Zero friction
Bonded

FIGURE 6.5

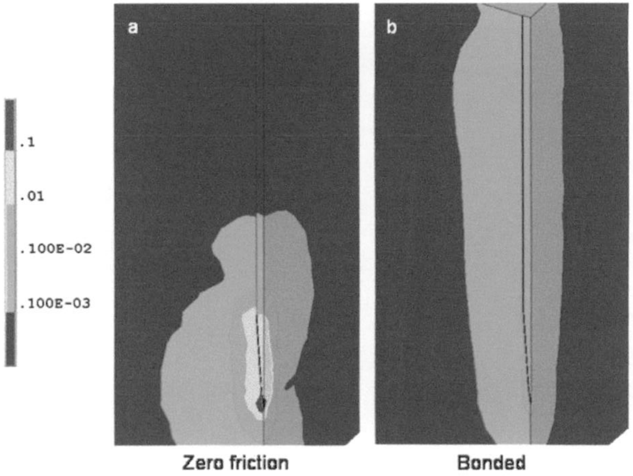

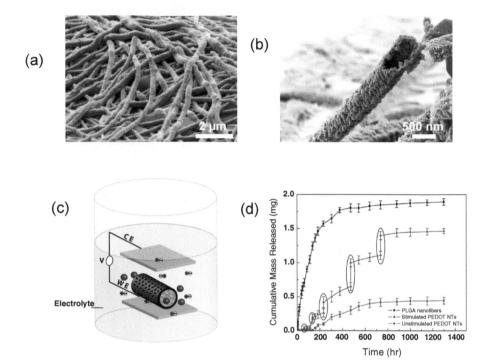

FIGURE 7.17

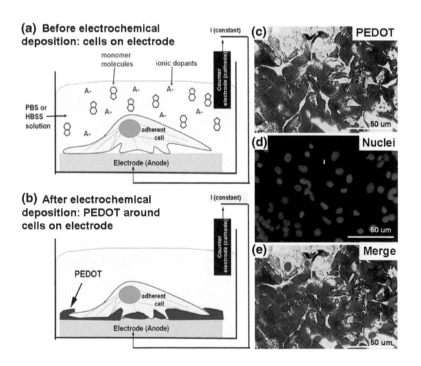

FIGURE 7.19

T - #0397 - 071024 - C43 - 234/156/13 - PB - 9780367387938 - Gloss Lamination